Environmental Science 4 Checkbook

M D W Pritchard

Butterworth Scientific
London Boston Durban Singapore Toronto Wellington

All rights reserved. No part of this publication may be reproduced or transmitted in any form or by any means, including photocopying and recording without the written permission of the copyright holder, application for which should be addressed to the publishers. Such written permission must also be obtained before any part of this publication is stored in a retrieval system of any nature.

This book is sold subject to the Standard Conditions of Sale of Net Books and may not be resold in the UK below the net price given by the Publishers in their current price list.

First published 1982

© M D W Pritchard 1982

British Library Cataloguing in Publication Data

Pritchard, M.D.W.
   Environmental science 4 checkbook.
   1. Human ecology
   I. Title
   333.7'0246   GF41

ISBN 0-408-00663-3
ISBN 0-408-00608-0 Pbk

Typeset by Scribe Design, Gillingham, Kent
Printed in England by Hartnoll Print Ltd, Bodmin, Cornwall

Contents

Preface vii

1 Environmental conditions and comfort 1
Air temperature 1
Mean radiant temperature 1
Air velocity 2
Saturation vapour pressure 2
Methods of specifying the humidity of the air 3
Measurement of relative humidity 5
Psychrometric chart 6
Comfort 8
Thermal indices 8
Comfort zones 9
Exercises 10

2 Steady state heat transfer 12
Thermal conductivity 12
Thermal resistivity 13
Conduction of heat 13
Thermal conductance and resistance 13
Surface resistance and resistance of air spaces 15
Radiation exchange 15
Convection exchange 15
Internal surface resistance 16
External surface resistance 16
Resistance of air spaces 17
Total resistance of a construction 17
Thermal transmittance (U-value) 18
Heat loss rate by ventilation 21
Steady state heat loss during winter heating 22
Composite constructions and heat bridges 26
Regulations relating to thermal insulation 27
Heating fuels 32
Temperature distribution 33
Pattern staining 35
Exercises 36

3 Condensation 40
Prediction of surface condensation 40
Prediction of interstitial condensation 42
Exercises 48

4 Solar and casual heat gains 50
Solar gains 50
Properties of thermal and solar radiation 50
Components of solar radiation 51
Transmission characteristics of glass 53
Methods of reducing solar gain through windows 56
Casual gains 57

Human bodies 57
Lighting 58
Exercises 58

5 Sound propagation and units of measurement 60
Sound waves 60
Velocity of sound in air 61
Velocity of sound in liquids and solids 62
Velocity, frequency and wavelength 62
Sound power, sound intensity and sound pressure 62
Decibel scales 64
Addition and subtraction of decibels 65
Loudness of pure tones 67
Sound level meters and weighting scales 68
Sound spectra 69
Exercises 70

6 Sound in rooms 71
Sound pressure level in a room 71
Sound absorbing materials 73
Reverberation time and Sabine's formula 74
Optimum reverberation time 75
Eyring's formula for reverberation time 76
Design of room shape 77
Exercises 78

7 Sound insulation 81
Airborne sound insulation 81
Sound reduction index of single leaf partitions 82
Composite partitions 85
Flanking transmission 87
Multiple leaf partitions 87
Measurement of airborne sound insulation 88
Prediction of sound levels through partitions 88
Impact sound insulation 90
Measurement of impact sound insulation of floors 90
Requirements for sound insulation 91
Rating sound insulation in buildings 93
Exercises 94

8 Hearing and the noise environment 97
The hearing mechanism 97
Hearing loss and damage 98
Code of practice for reducing the exposure of employed persons to noise 99
Noise acceptability and annoyance 101
Speech interference level (SIL) 101
Noise rating curves (NR) 102
Rating of industrial noise 104
Noise from construction and demolition sites 106
Traffic noise and dwellings 107
Perceived noise level and aircraft noise 109
Exercises 111

9 Noise control 114
Noise control at source 114

Noise control in transmission outdoors 115
Noise control indoors 118
Noise control at the receiver 119
Exercises 119

10 Vision, lighting units, colour and light sources 121
The eye and mechanism of vision 121
Light 121
Lighting units 122
Types of electric lamp 125
Colour of light sources 127
Surface colours 130
Colour rendering 131
Exercises 132

11 Lighting calculations 134
Direct component of illuminance due to a point source 134
Direct component of illuminance due to a linear source 137
Lumen method 141
Glare 147
Glare index tables 148
British zonal system (BZ) 151
Requirements of good lighting 152
Exercises 153

12 Daylighting 156
The overcast sky 156
Daylight factor and its components 156
Prediction of sky component 158
BRS sky component table 158
BRS daylight protractors 162
Prediction of externally reflected component 166
Internally reflected component 168
BRS inter-reflection formula 168
BRS nomograms 171
Calculation of total daylight factor 172
Exercises 179

Answers to exercises 184

Index 187

Note to Reader

As textbooks become more expensive, authors are often asked to reduce the number of worked and unworked problems, examples and case studies. This may reduce costs, but it can be at the expense of practical work which gives point to the theory.

Checkbooks if anything lean the other way. They let problem-solving establish and exemplify the theory contained in technician syllabuses. The Checkbook reader can gain *real* understanding through seeing problems solved and through solving problems himself.

Checkbooks do not supplant fuller textbooks, but rather supplement them with an alternative emphasis and an ample provision of worked and unworked problems. The brief outline of essential data—definitions, formulae, laws, regulations, codes of practice, standards, conventions, procedures, etc—will be a useful introduction to a course and a valuable aid to revision. Short-answer and multi-choice problems are a valuable feature of many Checkbooks, together with conventional problems and answers.

Checkbook authors are carefully selected. Most are experienced and successful technical writers; all are experts in their own subjects; but a more important qualification still is their ability to demonstrate and teach the solution of problems in their particular branch of technology, mathematics or science.

Authors, General Editors and Publishers are partners in this major low-priced series whose essence is captured by the Checkbook symbol of a question or problem 'checked' by a tick for correct solution.

Preface

The aim of this textbook is to provide information and assistance to students studying the TEC unit Environmental Science 4 on Building Courses. However, it will also prove valuable to students on other courses in Building and Architecture.

It is hoped that the combination of worked examples, facts and problems will enable the reader to grasp the essential information and appreciate its application in the design and construction of buildings. Many of the examples have been chosen to show the application of the information and to assist the reader in future studies. Each chapter includes a set of exercises for the student to work out with answers given at the end of the book.

The author is indebted to Colin Bassett, the Series editor of the Building Checkbooks, for his help and encouragement throughout the preparation of this work.

Michael Pritchard
Guildford County College of Technology

Acknowledgements

The author would like to express his gratitude to the following:

Chartered Institution of Building Services, Delta House, 222 Balham High Road, London SW12 9BS for permission to reproduce extracts from the CIBS Guide. CIBS publications are available from the above address.

Thorn Lighting, Commercial House, Lawrence Road, London, N15 4EG for their permission to reproduce the photometric data in chapter 10.

Building Research Establishment, Garston, Watford, WD2 7JR for permission to reproduce information on daylighting from BRE digests 41 and 42. BRE publications are available from the above address.

British Standards Institution, 2 Park Street, London W1A 2BS for permission to reproduce extracts from British Standards. Complete copies of the Standards can be obtained from the above address.

Her Majesty's Stationery Office for permission to reproduce extracts from the Building Regulations. Building Regulations can be obtained from HMSO or through booksellers.

Butterworths Technical and Scientific Checkbooks

General Editor for Building, Civil Engineering, Surveying and Architectural titles:
Colin R. Bassett, lately of Guildford County College of Technology.

General Editors for Science, Engineering and Mathematics titles:
J.O. Bird and A.J.C. May, Highbury College of Technology, Portsmouth.

A comprehensive range of Checkbooks will be available to cover the major syllabus areas of the TEC, SCOTEC and similar examining authorities. A comprehensive list is given below and classified according to levels.

Level 1 (Red covers)
Mathematics
Physical Science
Physics
Construction Drawing
Construction Technology
Microelectronic Systems
Engineering Drawing
Workshop Processes & Materials

Level 2 (Blue covers)
Mathematics
Chemistry
Physics
Building Science and Materials
Construction Technology
Electrical & Electronic Applications
Electrical & Electronic Principles
Electronics
Microelectronic Systems
Engineering Drawing
Engineering Science
Manufacturing Technology
Digital Techniques
Motor Vehicle Science

Level 3 (Yellow covers)
Mathematics
Chemistry
Building Measurement
Construction Technology
Environmental Science
Electrical Principles
Electronics
Microelectronic Systems
Electrical Science
Mechanical Science
Engineering Mathematics & Science
Engineering Science
Engineering Design
Manufacturing Technology
Motor Vehicle Science
Light Current Applications

Level 4 (Green covers)
Mathematics
Building Law
Building Services & Equipment
Construction Technology
Construction Site Studies
Concrete Technology
Economics for the Construction Industry
Geotechnics
Engineering Instrumentation & Control

Level 5
Building Services & Equipment
Construction Technology
Manufacturing Technology

1 Environmental conditions and comfort

In the discussion of comfort conditions and also heat transfer the following factors are important — air temperature, mean radiant temperature, air velocity and moisture content of the air. Each of these will be considered in turn.

Air temperature

The air temperature is most commonly measured using a mercury in glass thermometer. In order to obtain an accurate reading the thermometer bulb should be surrounded by a polished metal cylindrical shield to screen the bulb from radiant heat. If a current of air is passed over the bulb the thermometer reaches equilibrium more rapidly. Thermistors and thermocouples can also be used to measure air temperature.

Mean radiant temperature

The mean radiant temperature is a measure of the heat exchanges by radiation between a body and its surroundings. The mean radiant temperature is defined as the surface temperature of a uniform enclosure which will give the same net radiation exchange as the actual surfaces and their associated temperatures. In the centre of a room with unheated surfaces, the mean radiant temperature can be taken as:

$$\frac{\Sigma A_i t_i}{\Sigma A_i}$$

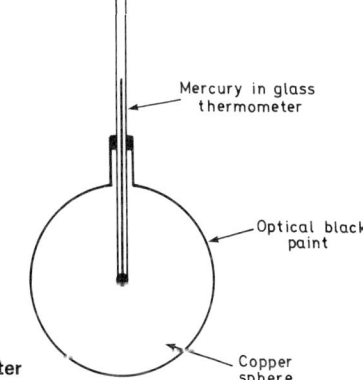

Fig 1 Globe thermometer

where A_i and t_i refer to the areas and temperatures of the individual surfaces.

The mean radiant temperature may be measured using a globe thermometer, which is illustrated in *Fig 1*. The mean radiant temperature is given by:

$(t_r + 273)^4 = (t_g + 273)^4 + 2.486 \times 10^8 \sqrt{v} \, (t_g - t_a)$

where t_r, t_g and t_a are the mean radiant, globe and air temperatures respectively and v is the air velocity. This equation is derived by considering the convection and radiation exchanges between the globe and its surroundings.

1

Problem 1 A mercury in glass thermometer shows an air temperature of 18°C and the globe temperature is 20°C. If the air velocity is 0.1 m/s calculate the mean radiant temperature.

From above

$(t_r + 273)^4 = (t_g + 273)^4 + 2.486 \times 10^8 \sqrt{v} \, (t_g - t_a)$
$(t_r + 273)^4 = (20 + 273)^4 + 2.486 \times 10^8 \sqrt{0.1} \, (20 - 18)$
$ = 73.7 \times 10^8 + 1.57 \times 10^8 = 75.27 \times 10^8$
$t_r + 273 = 294.5$
$t_r = 21.5°C$

Air velocity

The air velocity is important in comfort conditions and several methods exist for its measurement. One possible method is the hot wire anemometer in which a resistance wire is heated by an electric current; the temperature of the wire depends upon the air velocity over it.

Atmospheric moisture

Psychrometry, often referred to as hygrometry, is the study of mixtures of air and water vapour and an understanding of the physical processes is essential.

Saturation vapour pressure

Consider a container, as shown in *Fig 2(a)*, of fixed volume and maintained at a constant temperature $T°C$. Suppose, initially, that all the air is removed and the pressure is thus zero. If a small quantity of water is introduced this will evaporate to occupy the complete volume of the container as shown in *Fig 2(b)*.

The molecules of water vapour will create a vapour pressure denoted by e. As more water is introduced, the evaporation of this water causes the vapour pressure to

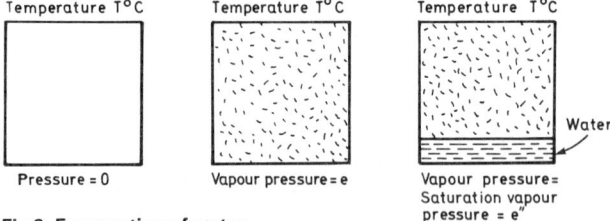

Fig 2 Evaporation of water

increase until a maximum vapour pressure is reached. This vapour is then said to be saturated and the vapour pressure is called the **saturation vapour pressure (s.v.p.)** and will be denoted by e''. If more water is introduced it will not evaporate but will exist as water, as shown in *Fig 2(c)*. If the temperature of the container is raised then more water will evaporate and the pressure in the container will increase. Thus the s.v.p. increases with temperature. Values of the s.v.p. are given in *Table 1* for a range of temperatures.

TABLE 1 Saturation vapour pressure over water

Temperature (°C)	s.v.p. (kPa)
0	0.61078
2	0.70547
4	0.81294
6	0.93465
8	1.0722
10	1.2272
12	1.4017
14	1.5977
16	1.8173
18	2.0630
20	2.3373
22	2.6430
24	2.9831
26	3.3608
28	3.7796
30	4.2430

It is sometimes useful to have an equation relating the s.v.p. to temperature. The following formula is adequate for most purposes:

$$\log_{10} e'' = \frac{7.5t}{237.3 + t} - 0.21429$$

where e'' is the s.v.p. in kPa and t is the temperature in °C.

Values of s.v.p. can be found from a psychrometric chart which will be introduced in a later section.

In the above discussion the container was initially evacuated but the water would evaporate in exactly the same manner if the container was originally filled with air. Dalton's law of partial pressures states that a mixture of water vapour and air behaves as if each constituent occupied the whole container at the same temperature. More specifically Dalton's law states that the total pressure is the sum of the air pressure and the water vapour pressure. Hence the mixture can be regarded as composed of dry air and water vapour acting independently. The above assumes that Dalton's law applies accurately to vapours, although this is not strictly true, the divergence under normal atmospheric conditions is negligible.

METHODS OF SPECIFYING THE HUMIDITY OF THE AIR

Water vapour in the air is not normally saturated and the following definitions and formulae are in accordance with B.S.1339 : 1965.

(i) **Absolute humidity**, d_v, is the mass of water vapour present in unit volume of moist air and is expressed in kg/m³. The absolute humidity is related in the vapour pressure, e, in kPa and the temperature, t, in °C by:

$$d_v = \frac{2.17e}{t + 273.15}$$

(ii) **Specific humidity**, q, is the mass of water vapour present in unit mass of moist air and is expressed in kg/kg moist air.

(iii) **Mixing ratio**, r, is the ratio of the mass of water vapour to the mass of dry air with which the water vapour is associated and is expressed in kg/kg dry air. The mixing ratio is related to the vapour pressure, e, in kPa and the total atmospheric pressure, P, in kPa by the equation:

$$r = \frac{0.622e}{P - e}$$

Note that the standard atmospheric pressure is 101.325 kPa.

(iv) **Relative humidity**, U_w, compares the amount of water vapour present in the air to the amount of water vapour present when saturated at the same temperature, thus:

$$U_w = \frac{\text{actual vapour pressure}}{\text{s.v.p. at the same temperature}} \times 100\%$$

$$= \frac{e}{e''} \times 100\%$$

(v) **Percentage saturation** is defined as:

$$\frac{\text{mixing ratio}}{\text{mixing ratio at saturation at the same temperature and pressure}} \times 100\%$$

If the mixture behaved as a perfect gas the percentage saturation and the relative humidity would be the same. Under normal conditions the difference is small and seldom exceeds 2%. The percentage saturation is related to the relative humidity by:

$$\frac{\text{percentage saturation}}{\text{relative humidity}} = \frac{P - e''}{P - e}$$

where e and e'' are the vapour pressure and s.v.p. at the same temperature respectively and P is the total atmospheric pressure.

(vi) **Dew point temperature**, t_d, is the temperature at which the water vapour pressure is equal to the s.v.p. If the temperature is reduced below this condensation will occur.

Problem 2 A mixture of air and water vapour at 16°C under an atmospheric pressure of 101.3 kPa has a water vapour pressure of 1.2 kPa.
(a) Determine:
  (i) the relative humidity,   (iii) the absolute humidity;
  (ii) the mixing ratio;   (iv) the percentage saturation.
(b) Find the relative humidity if the temperature of the mixture is reduced to 12°C.
(c) Determine the dew point temperature.

(a) (i) The vapour pressure e = 1.2 kPa
From *Table 1* the s.v.p. at 16°C is seen to be: e'' = 1.8173 kPa The relative humidity is defined as:

$$U_w = \frac{e}{e''} \times 100\% = \frac{1.2}{1.8173} \times 100\% = \mathbf{66.03\%}$$

(ii) From the formula given above the mixing ratio is:

$$r = \frac{0.622e}{P - e}$$

$$= \frac{0.622 \times 1.2}{101.3 - 1.2} = \mathbf{0.007457 \text{ kg/kg dry air}}$$

(iii) The absolute humidity can be calculated as follows:
$$d_v = \frac{2.17e}{t + 273.15} = \frac{2.17 \times 1.2}{16 + 273.15} = \mathbf{0.009 \text{ kg/m}^3}$$

(iv) From the above formula and the result of (a) (i):

$$\frac{\text{percentage saturation}}{\text{relative humidity}} = \frac{P - e''}{P - e}$$

$$\frac{\text{percentage saturation}}{66.03} = \frac{101.3 - 1.8173}{101.3 - 1.2} = 0.9938$$

percentage saturation = 66.03 × 0.9938 = **65.62**

It will be seen that the relative humidity and percentage saturation differ by less than 1%.

(b) From *Table 1*, s.v.p. at 12°C is 1.4017 kPa. Thus following (a) (i) above:

$$U_w = \frac{1.2}{1.4017} \times 100 = \mathbf{85.61\%}$$

Note that the vapour pressure remains unchanged but that the s.v.p. decreases with temperature with a consequent increase in the relative humidity.

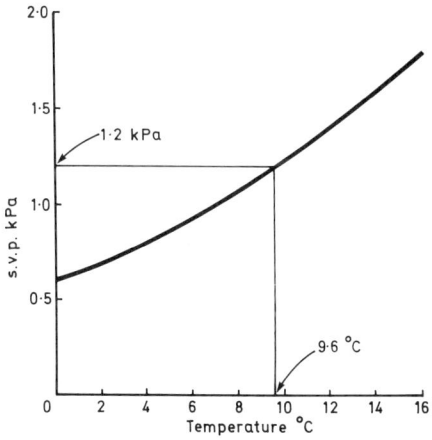

Fig 3 Problem 2: Dew point temperature

(c) The dew point temperature is the temperature at which the s.v.p. is equal to the vapour pressure of 1.2 kPa. An inspection of *Table 1* reveals that the s.v.p. at 10°C is 1.2272 kPa, thus the dew point temperature is just below 10°C. A closer estimate can be found by plotting a graph of the values in *Table 1* and reading the required temperature as 9.6°C, this is illustrated in *Fig 3*.

MEASUREMENT OF RELATIVE HUMIDITY

The wet and dry bulb hygrometer is a common instrument for measuring relative humidity. As water evaporates from the wick around the wet bulb the removal of latent heat reduces the temperature of the wet bulb below that of the dry bulb. The rate of evaporation, and hence the wet bulb temperature, depends on the water vapour pressure in the surrounding air. The wet and dry bulb readings thus enable the relative humidity to be found. This is often done using suitable tables or charts, but it is sometimes useful to calculate the relative humidity.

The actual vapour pressure is calculated from the wet bulb and dry bulb temperatures using one of the following formulae:

$e = e' - 6.66 \times 10^{-4} P(t - t')$
or $e = e' - 7.99 \times 10^{-4} P(t - t')$
where e = vapour pressure in kPa;
e' = s.v.p. at the wet bulb temperature. This can be found from *Table 1* or the previously given formula;
P = atmospheric pressure in kPa;
t = dry bulb temperature in °C;
t' = wet bulb temperature in °C.

The first formula is used for forced-ventilated hygrometers of the sling or Assmann type and the second formula is for hygrometers in a screen. An Assmann hygrometer is shown in *Fig 4*.

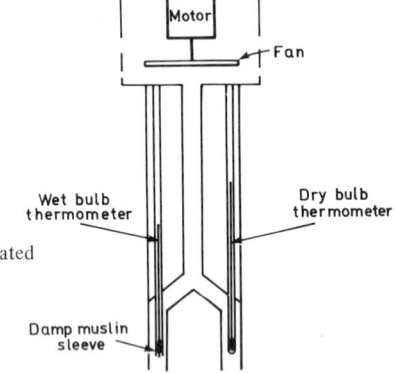

Fig 4 Assmann hygrometer

Problem 3 An Assmann hygrometer gave wet and dry bulb readings of 14°C and 18°C respectively. Assuming the atmospheric pressure to be 101 kPa determine the relative humidity of the air.

From *Table 1*:
s.v.p. at wet bulb temperature $e' = 1.5977$ kPa
s.v.p. at dry bulb temperature $e'' = 2.0630$ kPa
Since the hygrometer is forced-ventilated the actual vapour pressure is calculated using:
$e = e' - 6.66 \times 10^{-4} P(t - t')$
$e = 1.5977 - 6.66 \times 10^{-4} \times 101 (18 - 14)$
$= 1.3286$ kPa
The relative humidity $U_w = \dfrac{e}{e''} \times 100\% = \dfrac{1.3286}{2.0630} \times 100 =$ **64.4%**

PSYCHROMETRIC CHART

Many of the previous problems on humidity can be rapidly solved with the aid of the simplified psychrometric chart shown in *Fig 5*. The following examples should illustrate the use of this chart.

Problem 4 A mixture of air and water vapour at 16°C has a water vapour pressure of 1.2 kPa.
(a) Determine: (i) the relative humidity; (ii) the mixing ratio.
(b) Find the relative humidity if the temperature of the mixture is reduced to 12°C.
(c) Determine the dew point temperature.

(a) (i) The initial conditions of a temperature of 16°C and a vapour pressure of 1.2 kPa are shown on the chart in *Fig 5* by the point A. This point is between the

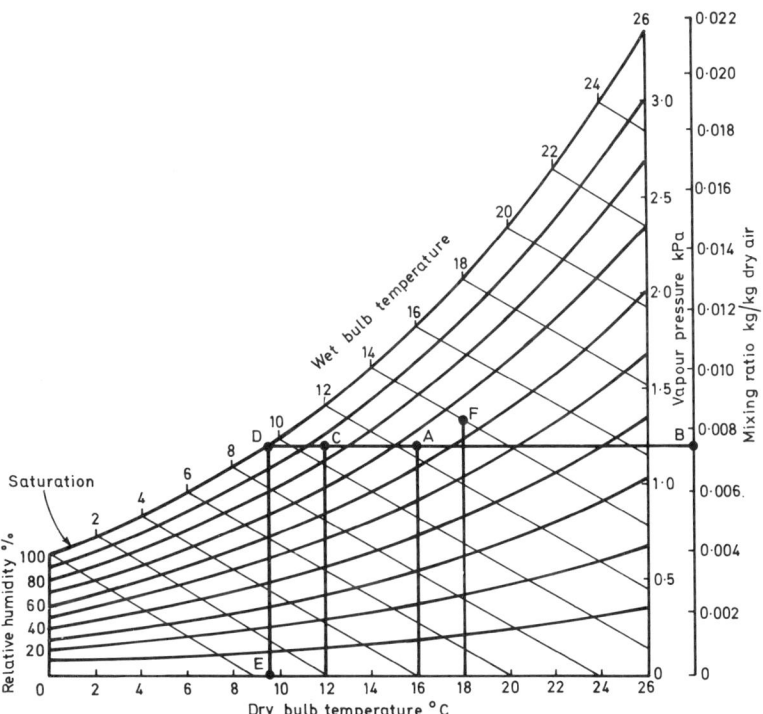

Fig 5 Psychrometric chart for Problems 4 and 5

60% and 70% curves of relative humidity. **The relative humidity can be estimated as 66%**.

(ii) The two right hand scales allow conversion between vapour pressure and mixing ratio. The conversion of the vapour pressure of 1.2 kPa is the point B on *Fig 5* and can be estimated as **0.0075 kg/kg dry air**.

(b) If the temperature is reduced to 12°C, the moisture content of the air being unchanged, the state of the mixture of air and water vapour is shown by the point C on *Fig 5*. This is between the 80% and 90% relative humidity curves and the relative humidity can be estimated as **85%**.

(c) If the mixture of air and water vapour is cooled further the curve of saturation is reached at point D on *Fig 5*. The corresponding dry bulb temperature, read at point E, is **9.6°C**.

The reader is advised to compare these answers with those obtained in *Problem 2*.

Problem 5 The wet and dry bulb readings of an Assmann hygrometer were 14°C and 18°C respectively. Using the psychrometer chart in *Fig 5* determine the relative humidity.

The lines for the wet bulb temperatures are those sloping downwards to the right across the chart. Following the line of 14°C wet bulb temperature until it meets the vertical line of 18°C dry bulb temperature leads to the point F on *Fig 5*.

The relative humidity is between the 60% and 70% curves and can be estimated as 65%. This compares favourably with the result obtained in *Problem 3*.

COMFORT

The human body continuously produces heat. For a sedentary adult under normal conditions this is about 110 W of which about 75% is lost by convection and radiation and about 25% by evaporation. These percentages will depend upon the environment; the percentage lost by evaporation rising as the air temperature rises due to the onset of sweating. The amount of heat produced by the human body also depends on the activity and will rise to over 400W for hard physical labour.

The provision of an environment in which the appropriate heat loss from the body can take place will not necessarily give a comfortable environment. By the use of physiological mechanisms, such as shivering and sweating, together with adjustments of activity and clothing the human is capable of maintaining a constant body temperature in a wide range of environments. The comfort zone differs from person to person and depends on activity, clothing, age, sex, season, and degree of acclimatisation.

Thermal indices

The basic factors affecting thermal comfort are: air temperature, mean radiant temperature, air velocity and relative humidity. Many attempts have been made to devise a single index which combines some or all of these factors. The following are but a few of these attempts.

(i) *Air temperature*. This has been widely used but is inadequate when there is a combination of extreme temperature and high air velocity or a strong exchange of radiation between a person and their surroundings.

(ii) *Equivalent temperature*. This was originally the reading taken with an eupatheoscope which was an electrically heated cylinder, the energy supply to which was regulated so that its surface temperature was similar to that of a clothed human body. The equivalent temperature was not appropriate when humidity effects were important.

(iii) *Globe temperature*. The globe thermometer has been previously described as a method of measuring the mean radiant temperature. The previously given equation for the globe thermometer may be simplified under normal conditions to:

$$t_g = \frac{t_r + 2.35\, t_a \sqrt{v}}{1 + 2.35 \sqrt{v}}$$

where t_g, t_r and t_a are the globe, mean radiant and air temperatures respectively and v is the air velocity.

As a thermal index it does not take account of relative humidity but it correlates well with subjective assessments of warmth in buildings.

(iv) *Dry resultant temperature*. This is the temperature at the centre of a blackened globe 100 mm in diameter and an equation similar to that for the globe thermometer relates the dry resultant temperature, the air temperature, the mean radiant temperature and the air velocity. A simplified form of the equation for normal conditions in buildings is:

$$t_c = \frac{t_r + t_a\sqrt{10v}}{1 + \sqrt{10v}}$$

where t_c, t_r, t_a are the dry resultant, mean radiant and air temperatures respectively.

Since the globe used is smaller the radiant component has less effect on the reading than with the 150 mm diameter globe thermometer.

At a velocity of 0.1 m/s, which is typical inside a building it will be seen that:

$$t_c = \frac{1}{2}t_r + \frac{1}{2}t_a$$

(v) environmental temperature: though not intended as a thermal index this temperature is of considerable importance. For a uniform cubical enclosure with one external wall the environmental temperature, t_{ei}, at the centre of the room is given by:

$$t_{ei} = \frac{2}{3}t_r + \frac{1}{3}t_a$$

Problem 6 In a room the air temperature is 22°C and the mean radiant temperature is 18°C. Evaluate the dry resultant temperature and the environmental temperature for air velocities of (i) 0.1 m/s and (ii) 0.3 m/s.

In each case

$$t_c = \frac{t_r + t_a\sqrt{10v}}{1 + \sqrt{10v}} \quad \text{and} \quad t_{ei} = \frac{2}{3}t_r + \frac{1}{3}t_{ei}$$

where in this problem $t_a = 22°C$ and $t_r = 18°C$.

(i) The dry resultant temperature is found as:

$$t_c = \frac{18 + 22\sqrt{10 \times 0.1}}{1 + \sqrt{10 \times 0.1}} = \mathbf{20°C}$$

The environmental temperature is:

$$t_{ei} = \frac{2}{3} \times 22 + \frac{1}{3} \times 18 = \mathbf{20.7°C}$$

(ii) $t_c = \dfrac{18 + 22\sqrt{10 \times 0.3}}{1 + \sqrt{10 \times 0.3}} = \mathbf{20.5°C}$

The environmental temperature is unchanged, thus $t_{ei} = \mathbf{20.7°C}$.

Comfort zones

The CIBS Guide (Section A1) states that the majority of people will be neither warm nor cool in winter in rooms where the dry resultant temperature is between 19°C and 23°C when the air speed is less than 0.1 m/s. This applies for sedentary occupations.

Humidity will have no effect on the warmth as long as the relative humidity is between 40% and 70% and the dry resultant temperature is near the preferred value. The Guide gives a table of recommended design values for dry resultant temperature for a wide range of building uses.

A number of additional points need to be considered in applying the above recommendations, i.e.:

1. If physical activity is being undertaken the value of the dry resultant temperature may be reduced by up to 3 to 5°C.

2. For air speeds in excess of 0.1 m/s an increase in dry resultant temperature will be necessary although it is likely that velocities in excess of 0.3 m/s will prove unacceptable in winter.

3. If the occupants can be persuaded to wear extra clothing it is possible to reduce the resultant temperature but the effect is limited. Extra clothing will keep the body warm but will leave the extremities unprotected and uncomfortable. Also the range of activity which is possible is reduced.

4. Although it is unlikely that all people will be satisfied at the design dry resultant temperature the percentage of dissatisfied people will not increase significantly if the temperature is within 1.5°C of the optimum temperature. Temperature variations throughout the room are thus important if they exceed this value, such variations can occur due to variations in air temperature or radiant temperature. It is preferable that the air temperature is higher at foot level than head level. Apart from positions near to

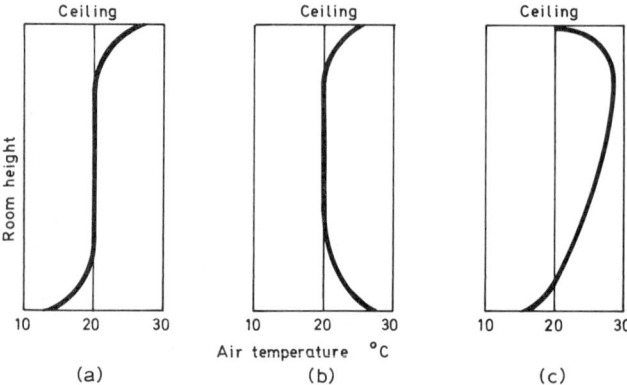

Fig 6 Vertical temperature gradients in a room with different heating systems (a) wall mounted radiator; (b) underfloor heating; (c) warm air with low level inlet

a window the air temperature gradient is mainly vertical. The temperature depends on the size and position of the heating unit. *Fig 6* shows some typical cases.

5. Variations of radiant temperature may cause discomfort particularly if the radiation is different from different directions. Local cooling can occur near cold surfaces such as windows. Local heating can occur near hot surfaces or radiant heating sources. Direct solar radiation through windows may increase the radiant temperature. The CIBS Guide gives guidance for these situations.

EXERCISES (answers on page 184)

1. The air and globe temperatures in a room are 22°C and 20°C respectively. Evaluate the mean radiant temperature for an air velocity of (i) 0.1 m/s and (ii) 0 m/s.

2. A mixture of air and water vapour at 16°C has a mixing ratio of 0.009 kg/kg dry air. If the atmospheric pressure is 101.3 kPa calculate:
   (i) the water vapour pressure; (ii) the relative humidity;
   (iii) relative humidity if the temperature is reduced to 14°C.

3. A mixture of air and water vapour in a room had a temperature of 18°C and a relative humidity of 60%. Using the psychrometric chart determine:
   (i) the water vapour pressure,
   (ii) the mixing ratio;
   (iii) the relative humidity if the temperature is reduced to 14°C;
   (iv) the dew point temperature.

4. It was noted that condensation began to occur on the inside of a window when its surface temperature was 7°C. The air temperature in the room was 18°C. Find the relative humidity of the air in the room.

5. The wet and dry bulb temperatures recorded by an Assmann hygrometer were 20°C and 16°C respectively. Determine the relative humidity and dew point temperature of the air using the psychrometric chart.

In the following exercises select the correct options

6. In a room, where the air velocity is about 0.1 m/s, the air temperature is 18°C and the mean radiant temperature is 20°C.
   (a) the dry resultant temperature is greater than the air temperature.
   (b) the dry resultant temperature is less than the air temperature.
   (c) the environmental temperature is greater than the dry resultant temperature.
   (d) the environmental temperature is less than the dry resultant temperature.

7. For sedentary work in rooms in winter the dry resultant temperature should be:
   (a) less than 19°C;
   (b) over 23°C;
   (c) between 19°C and 23°C;
   (d) between 13°C and 20°C.

8. For sedentary work in rooms in winter the air speed should be:
   (a) up to 1 m/s; (b) up to 0.1 m/s; (c) above 0.3 m/s.

9. The occupants of a building will be unaffected by relative humidity, assuming the temperature is suitable, if it is:
   (a) less than 40%,
   (b) between 10–90%;
   (c) above 70%;
   (d) between 40–70%.

2 Steady state heat transfer

THERMAL CONDUCTIVITY (λ)

The ability of a material to conduct heat is called its thermal conductivity and denoted by the symbol λ. The thermal conductivity is defined as the quantitiy of heat conducted through one metre thickness of material having a cross-sectional area of one square metre, in one second if a temperature difference of $1°C$ is maintained between the faces. The units are W/mK.

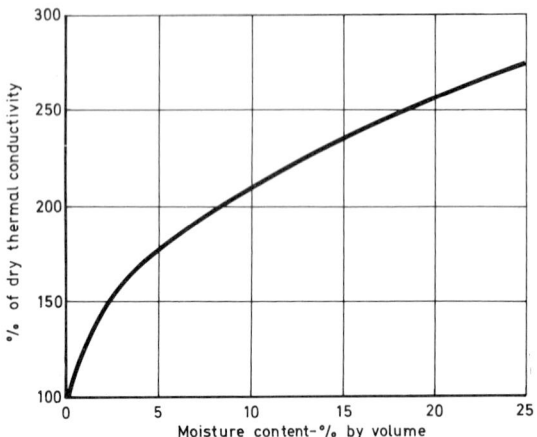

Fig 1 Variation of thermal conductivity with moisture content

The values of conductivity vary between 0.03 W/mK for good insulating materials and 400 W/mK for metals. The conductivity of air is about 0.025 W/mK which is lower than all solid materials thus good insulating materials will have a porous structure.

Since the conductivity of water is considerably higher than that of air the conductivity of a porous material will increase if it becomes wet. The increase in conductivity with moisture content is shown in *Fig 1*.

Problem 1 The thermal conductivity of a certain brick when dry is 0.5 W/mK. Determine the conductivity of the brick when it is exposed in an external wall and has a moisture content of 5% by volume.

From *Fig 1* it will be seen that the conductivity is 175% of its dry value.
Thus $\lambda = 1.75 \times 0.5 = \mathbf{0.875\ W/mK}$.

It will be seen that the change in conductivity is large and care must be taken to use appropriate values for external masonry, suitable values of moisture content would be 1% and 3% for brickwork and concrete protected from rain and 5% for these materials if exposed to rain.

THERMAL RESISTIVITY (r)

The thermal resistivity is the reciprocal of the thermal conductivity and has units of mK/W. The resistivity is more convenient to use in some instances than the conductivity.

Problem 2 The thermal conductivity of a material is 2W/mK. Determine its resistivity.

Since $r = 1/\lambda$; $\therefore r = 1/2 = \mathbf{0.5\ mK/w}$

Conduction of heat

The heat flow rate, in watts, through the slab of material shown in *Fig 2* which has a cross sectional area of A square metres, thickness l m and face temperatures of t_1 °C and t_2 °C is given by

$$Q = \frac{\lambda A}{l}(t_1 - t_2)$$

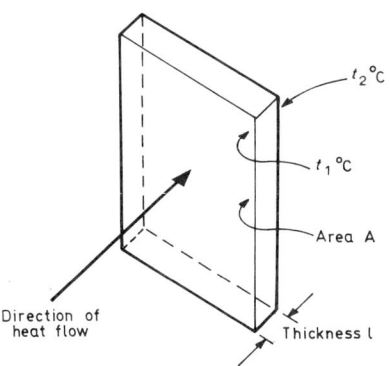

Fig 2 Heat conduction through a slab of material

Problem 3 Determine the heat flow rate through 4 m² of brickwork of thickness 0.11 m when the temperatures of the faces are 20°C and 10°C. The conductivity of the brick is 0.75 W/mK.

Using the formula: $Q = \frac{\lambda A}{l}(t_1 - t_2)$

$$Q = \frac{0.75 \times 4}{0.11}(20 - 10) = \mathbf{272.7\ W}$$

Thermal conductance and thermal resistance

In the heat conduction formula the physical properties of the material are contained in the term $\lambda A/l$, this quantity is called the thermal conductance. The reciprocal of this quantity, $l/\lambda A$ is termed the thermal resistance. It is usual to base calculations on constructions on an area of 1 m², thus $A = 1$ in the above definitions, hence:

thermal conductance = $\dfrac{\lambda}{l}$ W/m²K

thermal resistance $R = \dfrac{l}{\lambda}$ m²K/W

Problem 4 Calculate the thermal resistance of 0.1 m thickness of dry aerated concrete having a conductivity of 0.25 W/m²K. Determine the thermal resistance of this concrete if its moisture content becomes 3%.

From above: $R = l/\lambda = 0.1/0.25 = \mathbf{0.4\ m^2 K/W}$.
If the moisture content is 3%, it will be seen from *Fig 1* that, the conductivity becomes $1.6 \times 0.25 = 0.4$ W/MK.
Hence:
$R = l/\lambda = 0.1/0.4 = \mathbf{0.25\ m^2\ K/W}$
For the composite slab shown in *Fig 3* it may be shown that the total resistance R is given by:

$R = R_1 + R_2 + R_3 + \ldots$

where R_1, R_2, R_3 are the individual resistances of the components.

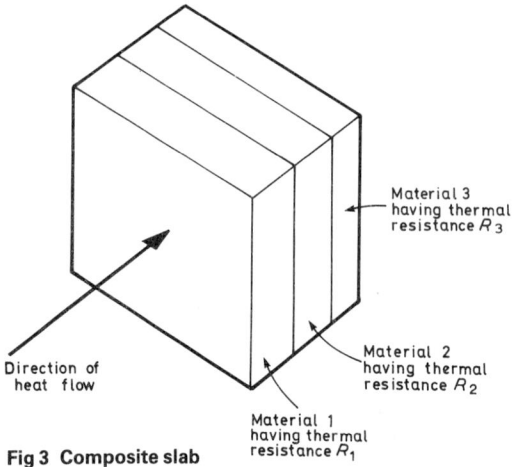

Fig 3 Composite slab

Material 1 having thermal resistance R_1

Material 2 having thermal resistance R_2

Material 3 having thermal resistance R_3

Direction of heat flow

Problem 5 A solid wall consists of 13 mm of plaster ($\lambda = 0.4$ W/mK), 220 mm of brickwork ($\lambda = 0.8$ W/mK) and is rendered externally with 20 mm of rendering ($\lambda = 0.9$ W/mK). Determine the total resistance of the structure.

For each material $R = l/\lambda$, in calculating the resistance care must be taken to express the thickness in metres. The total resistance is then found by summing the resistances, thus:

$R = \dfrac{0.013}{0.4} + \dfrac{0.220}{0.8} + \dfrac{0.020}{0.9} = \mathbf{0.33\ m^2 K/W}$

SURFACE RESISTANCE AND RESISTANCE OF AIR SPACES

The flow of heat through a structure is dependent not only on the thermal properties of the structure but also upon the heat exchanges between the surfaces and the surroundings. There are two mechanisms to be considered; these are radiation and convection exchanges.

Radiation exchange

The radiation transfer is governed by Stefan's Law which may be stated as:
$$q_r = \sigma E A (T_1^4 - T_2^4)$$
where q_r = heat transfer rate in W;
σ = Stefan's constant = 5.7×10^{-8} W/m²K;
A = area of surface;
T_1, T_2 = temperatures in degrees Kelvin of the surface and the surroundings;
E = emissivity factor of the surface.

For practical purposes in construction it is not convenient to work with the difference of the fourth powers of the temperature and the following approximation is used:
$$q_r = AEh_r (t_s - t_r)$$
where h_r is the radiation conductance, t_r and t_s are the mean surface temperature of the surroundings and the temperature of the surface respectively in °C.

Heat transfer is usually related to an area of one square metre so that:
$$q_r = Eh_r (t_s - t_r) \text{ W/m}^2$$

The values of the emissivity factor may be taken as 0.9 for most building materials, 0.2 for dull aluminium and 0.05 for polished aluminium. The values of h_r are temperature dependent and vary from 5.7 W/m²K at 20°C to 4.6 W/m²K at 0°C.

> *Problem 6* Determine the heat transfer rate by radiation from the surface of a sheet of plasterboard if the mean surface temperature of the surroundings is 2°C and its surface temperature is 10°C and (i) its emissivity factor is 0.9; (ii) if it is covered in aluminium foil with an emissivity factor of 0.2. Assume $h_r = 4.95$ W/m²K.

(i) From above:
$q_r = Eh_r (t_s - t_r)$
$q_r = 0.9 \times 4.95 (10 - 2) =$ **35.6 W/m²**
(ii) Using the same formula:
$q_r = 0.2 \times 4.95 (10 - 2) =$ **7.9 W/m²**

It will be noted that the heat transfer is considerably less and this accounts for the use of foil backed materials as insulation materials and serves to confirm the well known statement that dark materials are good radiators and shiny surfaces are poor radiators.

Convection exchange

The heat transfer rate at a surface is approximately given by:
$$q_c = Ah_c (t_s - t_a)$$
where A = area of surface;
t_s, t_a = temperatures of the surface and the surrounding air;
h_c = convection conductance.

As is usual an area of one square metre is considered so that:
$$q_c = h_c (t_s - t_a) \text{ W/m}^2$$
The value of h_c depends on whether there is air movement over the surface and also the direction of heat flow. For internal surfaces where there is no wind and free convection takes place the values of h_c that are usually employed are 3.0 W/m²K for walls, 4.3 W/m²K for upward heat flow into ceilings and 1.5 W/m²K for downward heat flow into floors. For external surfaces which are exposed to wind then h_c can be found using:
$$h_c = 5.8 + 4.1v \text{ for air velocities of less than 5 m/s}$$
and $h_c = 7.8v^{0.8}$ for air velocities in excess of 5 m/s
where v is the air velocity.

INTERNAL SURFACE RESISTANCE

The total heat transfer rate at an internal surface is the sum of the radiation and convection exchanges; thus:
$$q = q_r + q_c$$
$$q = Eh_r (t_r - t_s) + h_c (t_a - t_s) \text{ W/m}^2$$
The order of the temperatures are taken so that the heat transferred into the surface is positive. The use of two different temperatures, t_r and t_a makes practical calculations difficult and the above equation may be written as
$$q = (Eh_r + h_c)(t_{ei} - t_{si})$$
where t_{si} is the temperature of the inside surface, the i being included in the subscript to indicate an inside surface, and t_{ei} is defined as the inside environmental temperature.

The environmental temperature t_{ei} is a combination of the inside air temperature, t_{ai} and the inside mean radiant temperature t_{ri} and for normal cases can be found from:
$$t_{ei} = \frac{2}{3} t_{ri} + \frac{1}{3} t_{ai}$$

For practical convenience it is helpful to define the internal surface resistance R_{si} as:
$$R_{si} = \frac{1}{Eh_r + h_c} \text{ m}^2\text{K/W}$$
The heat flow rate may then be written as
$$q = \frac{1}{R_{si}} (t_{ei} - t_{si}) \text{ W/m}^2$$

EXTERNAL SURFACE RESISTANCE

In an analogous manner to the case of the internal surface resistance the external surface resistance is defined as
$$R_{so} = \frac{1}{Eh_r + h_c}, \text{ where appropriate values of } h_r \text{ and } h_c \text{ are taken.}$$
The heat flow rate to the outside may then be written as:
$$q = \frac{1}{R_{so}} (t_{so} - t_{eo})$$
where t_{so} and t_{eo} are the outside surface and environmental temperatures respectively.

Problem 7. Calculate the internal surface resistance for a wall:
(i) having a high emissivity factor ($E = 0.9$);
(ii) having a low emissivity factor ($E = 0.05$).
The appropriate values of conductance are $h_r = 5.7$ W/m²K and $h_c = 3.0$ W/m²K.

(i) From above $R_{si} = \dfrac{1}{Eh_r + h_c} = \dfrac{1}{0.9 \times 5.7 + 3.0} = 0.123$ m²K/W

(ii) Using $E = 0.05$ in the above formula:

$$R_{si} = \dfrac{1}{0.05 \times 5.7 + 3.0} = 0.304 \text{ m}^2\text{K/W}$$

The variation in surface resistance is large. More detailed values are given in the CIBS Guide (Section A3).

Problem 8 Calculate the outside surface resistance for a wall at a windspeed of (i) 2 m/s; (ii) 6 m/s if $E = 0.9$ and $h_r = 4.7$ W/m²K.

(i) It is necessary to calculate a value for h_c.
$h_c = 5.8 + 4.1v$ where v is the windspeed
$h_c = 5.8 + 4.1 \times 2 = 14$.
Then $R_{so} = \dfrac{1}{Eh_r + h_c} = \dfrac{1}{0.9 \times 4.7 + 14} = 0.055$ m² K/W

(ii) Using a windspeed of 6 m/s it is found that:
$h_c = 5.8 + 4.1 \times 6 = 30.4$

Then $R_{so} = \dfrac{1}{0.9 \times 4.7 + 30.4} = 0.029$ m² K/W

RESISTANCE OF AIR SPACES

The heat transfer across a cavity or airspace will take place by radiation, convection and conduction. The conduction component being in many cases the least important. The radiation transfer depends on the emissivity factor of the two surfaces of the cavity or airspace. It may be shown that the effective emissivity factor is given by:

$$\dfrac{E_1 E_2}{E_1 + E_2 - E_1 E_2}$$

where E_1 and E_2 are the emissivity factors of the two surfaces.

If both sides of the airspace are of high emissivity factor, $E_1 = E_2 = 0.9$, then the effective emissivity factor is 0.82. If one side is lined with reflective foil so that $E_1 = 0.9$ and $E_2 = 0.05$ then the effective emissivity factor is very nearly 0.05. Thus lining one side of the airspace with reflective foil dramatically reduces the effective emissivity factor.

The convection transfer will depend on the width of the airspace, its inclination, the direction of heat flow and amount of ventilation of the airspace. The resistance of the air space will increase as its width increases up to about 25 mm. A horizontal airspace presents a higher resistance to downward heat flow than to upward heat flow. Ventilation of an airspace will considerably reduce its resistance.

The result of these complex heat exchanges across an airspace are summarised by a single resistance R_a. For instance the airspace in cavity wall construction may be taken to have a resistance of 0.18 m² K/W

Total resistance of a construction

The total resistance of a construction can now be written as:
$$R = R_{si} + R_1 + R_2 + \ldots + R_a + R_{so}$$

> *Problem 9* A cavity wall consists of 0.013 m of plaster ($\lambda = 0.4$ W/mK) on 0.1 m of lightweight concrete blocks ($\lambda = 0.17$ W/mK); the outer leaf is 0.102 m of brickwork ($\lambda = 0.9$ W/mK). The inside, outside and cavity resistances are:
> $R_{si} = 0.123$ m²K/W, $R_{so} = 0.055$ m²K/W and $R_a = 0.18$ m²K/W
> Calculate the total thermal resistance of the construction.

Using the above formula:
$$R = 0.123 + \frac{0.013}{0.4} + \frac{0.1}{0.17} + 0.18 + \frac{0.102}{0.9} + 0.055$$
$$= 1.092 \text{ m}^2\text{K/W}$$

It will be noted that for the solid materials that the resistance in each case is calculated using $\frac{l}{\lambda}$ and that the resistances have been written down in order starting from the inside of the wall.

THERMAL TRANSMITTANCE (U-value)

This quantity is of the utmost importance in calculating heat loss from a building. The interpretation of this quantity can be arrived at by considering the heat conduction equation again:
$$Q = \frac{\lambda A}{l}(t_1 - t_2)$$

This can be written as
$$Q = \frac{A}{l/\lambda}(t_1 - t_2) = \frac{A}{R}(t_1 - t_2)$$
where R is the thermal resistance, $\frac{l}{\lambda}$.

For a composite construction R will be the total resistance and t_1 and t_2 will be the inside and outside environmental temperatures denoted by t_{ei} and t_{eo} respectively. Hence
$$Q = \frac{A}{R}(t_{ei} - t_{eo})$$

It has become conventional to put $U = \frac{1}{R}$ where U is called the thermal transmittance, thus:
$$Q_u = AU(t_{ei} - t_{eo})$$
where $U = \frac{1}{R}$ and Q_u is termed the fabric loss.

The thermal transmittance, usually referred to as the U-value, represents the heat flow rate in watts through one square metre of a construction when a temperature difference of 1°C exists between inside and outside environmental temperatures. By referring to the units of resistance it will be seen that the units of the U-value are W/m²K.

Since the surface resistances and air space resistances depend on the nature of the materials, the exposure of the building and the direction of heat flow the U-value will also depend on these variables. The CIBS Guide (Section A3) gives the U-values of many forms of construction for different exposures and orientations.

Problem 10 Calculate the U-value for the construction detailed in *Problem 9*.

Since $U = \frac{1}{R}$, where R is the total resistance,

The U-value is given by:

$$U = \frac{1}{1.092} = 0.916 \text{ W/m}^2\text{K}.$$

Problem 11 Calculate the U-value for a flat roof having the following properties:
19 mm asphalt ($\lambda = 0.5$ W/mK)
50 mm lightweight concrete screed ($\lambda = 0.42$ W/mK)
150 mm solid concrete ($\lambda = 1.4$ W/mK)
16 mm plaster ceiling ($\lambda = 0.5$ W/mK)
The inside and outside surface resistances may be taken as 0.106 and 0.045 m² k/W respectively.
Determine also the heat loss through 50 m² of this roof when the inside and outside environmental temperatures are 20°C and −1°C respectively.

The total resistance is found as:

$$R = 0.106 + \frac{0.016}{0.5} + \frac{0.150}{1.4} + \frac{0.050}{0.42} + \frac{0.019}{0.5} + 0.045$$

$$= 0.447 \text{ m}^2 \text{ k/W}$$

Since $U = \frac{1}{R}$ then $U = \frac{1}{0.447} = \textbf{2.24 W/m}^2\textbf{K}$

The heat loss is given by:

$$Q_u = AU(t_{ei} - t_{eo})$$

$$= 50 \times 2.24 (20 - (-1)) = 50 \times 2.24 \times 21 = \textbf{2352 W}$$

Problem 12 The U-value of conventional tiled roof construction is 1.5 W/m²K. Determine the thickness of fibreglass quilt that would be required to be laid between the ceiling joists to reduce the U-value to 0.5 W/m²K. The conductivity of the fibreglass is 0.035 W/mK.

In order to answer this problem it is necessary to deal with resistances rather than the U-values. By using the definition of the U-value it will be seen that $R = \frac{1}{U}$

Existing resistance $R = 1/1.5 = 0.667$ m²k/W
Required resistance $R = 1/0.5 = 2.0$ m²k/W.
Additional resistance to be provided by the firbeglass is $2.0 - 0.667 = 1.333$ m²k/W.

Knowing that the resistance of the fibreglass is given by $\frac{l}{\lambda}$ the following equation can be written:

$$\frac{l}{\lambda} = 1.333$$

$$\frac{l}{0.035} = 1.333$$

$l = 1.333 \times 0.035 = \mathbf{0.047\ m}$

The thickness of the fibreglass is 47 mm or 50 mm from a practical point of view. It is to be noted that the thermal resistance of the loft air space will not be changed by the addition of the fibreglass.

Problem 13 A detached bungalow, 10 m by 12 m on plan, has a ceiling height of 2.4 m. The external cavity walls have a U-value of 0.96 W/m²K. The external walls include 18 m² of single glazing with a U-value of 0.64 W/m²K. The pitched roof has a U-value of 1.5 W/m²K. The average internal and external environmental temperatures may be taken as 19°C and −1°C respectively. Determine:
(a) the heat loss rate through the fabric.
(b) the heat loss rate when 75 mm of vermiculite granules (λ = 0.065 W/mK) are placed between the ceiling joists.
(c) when the 59 mm cavity in the external walls is filled with foam (λ = 0.036 W/mK). The cavity fill is in addition to the vermiculite in (b). The thermal resistance of the cavity before filling is 0.18 m²K/W.

(a) Before proceeding to find the heat loss rate note that:
(i) in finding the area of the cavity walls allowance must be made for the area of the glazing:

area of cavity walls = 2(10 × 2.4 + 12 × 2.4) − 18 = 87.6 m²

(ii) U-values for pitched roofs are quoted so that the appropriate area to be used is the plan area.

The heat loss calculations can now be tabulated as in *Table 1*.

(b) It is necessary to calculate a new U-value for the roof construction. The addition of the vermiculite will not significantly alter the thermal resistance of the air space in a pitched roof construction; hence, the thermal resistance of the vermiculite can be added to the existing resistance.

TABLE 1

Element	Area (m²)	Heat loss rate $Q_u = AU(t_{ei} - t_{eo})$
Cavity walls	87.6	87.6 × 0.96 × (19 − (−1)) = 1682
Windows	18	18 × 5.6 (19 − (−1)) = 2016
Floor	120	120 × 0.64 × (19 − (−1)) = 1536
Roof	120	120 × 1.5 × (19 − (−1)) = 3600
		Total heat loss rate = **8834 W**

Existing resistance $R = \dfrac{1}{U} = \dfrac{1}{1.5} = 0.667 \text{ m}^2\text{K/W}$

Resistance of vermiculite $= \dfrac{l}{\lambda} = 0.075/0.065 = 1.154 \text{ m}^2\text{K/W}$

Total resistance $= 0.667 + 1.154 = 1.821 \text{ m}^2\text{K/W}$.

New U-value $= \dfrac{1}{1.821} = 0.55 \text{ W/m}^2\text{K}$.

Heat loss rate through the roof $= AU(t_{ei} - t_{eo})$
$= 120 \times 0.55 \times (19 - (-1)) = 1320 \text{ W}$.

The heat loss rate through the other elements is unchanged so that the total heat loss rate is now: $1682 + 2016 + 1536 + 1320 = \mathbf{6554 \text{ W}}$.

(c) When the cavity is filled the original resistance of the airspace is lost and is replaced by the resistance of the cavity fill.

Existing resistance $R \quad = \dfrac{1}{U} = \dfrac{1}{0.96} = 1.042 \text{ m}^2\text{K/W}$

Resistance of cavity fill $= \dfrac{l}{\lambda} = \dfrac{0.050}{0.036} = 1.389 \text{ m}^2\text{K/W}$

Resistance of air space to be deducted $= 0.18 \text{ m}^2\text{K/W}$

Total resistance $\quad = 1.042 + 1.389 - 0.18 = 2.251 \text{ m}^2\text{K/W}$

New U-value $\quad = \dfrac{1}{2.251} = 0.44 \text{ W/m}^2\text{K}$

Heat loss rate through the walls $= AU(t_{ei} - t_{eo})$
$= 87.6 \times 0.44 \times (19 - (-1)) = 771 \text{W}$

The total heat loss rate is now: $771 + 2016 + 1536 + 1320 = \mathbf{5643 \text{ W}}$

HEAT LOSS RATE BY VENTILATION

Assuming that the air required for ventilation is drawn from the outside the heat rate required to heat the ventilating air is given by:

$Q_v = C\rho NV(t_{ai} - t_{ao})$

where C = specific heat capacity of air in J/kgK;

ρ is the density of air in kg/m^3;

N is the ventilation rate in air changes per hour;

V is the volume of the room in m^3;

t_{ai}, t_{ao} are the inside and outside air temperatures.

Since the air in a building is a variable mixture of air and water vapour the density and specific heat capacity are variable; a typical figure for $C\rho$ in the above equation is 1200 J/m^{3}K. As the ventilation rate is in air changes per hour it is necessary to convert this to seconds by dividing by 3600.

Hence $Q_v = C\rho NV(t_{ai} - t_{ao})$
$= \dfrac{1200 \text{ NV}}{3600}(t_{ai} - t_{ao})$
$= \dfrac{1}{3}NV(t_{ai} - t_{ao})$

Problem 14 Determine the heat loss rate by ventilation from the bungalow in *Problem 13* if the ventilation rate is 1.5 air changes per hour and the inside and outside air temperatures are 19°C and −1°C respectively.

The volume of the bungalow = 10 × 12 × 2.4 = 288 m³
The heat loss rate by ventilation is given by:

$$Q_v = \frac{1}{3} NV (t_{ai} - t_{ao})$$
$$= \frac{1}{3} \times 1.5 \times 288 (19 - (-1))$$
$$= 2880 \text{ W}.$$

Steady state heat loss during winter heating

The total heat loss rate is the sum of the fabric loss, Q_u, and the ventilation loss, Q_v. The expressions for these losses are:

$$Q_u = AU (t_{ei} - t_{eo})$$
$$\text{and} \quad Q_v = \frac{1}{3} NV (t_{ai} - t_{ao})$$

It will be seen that the temperatures involved in each of these expressions are different. Under suitable circumstances it is possible to assume that the air temperatures and environmental temperatures are the same.

In winter in Great Britan, since the sky is overcast for long periods of time it is usual to assume that the outside air and environmental temperatures are the same. This assumption is not true when considering summertime conditions. For well insulated buildings the temperatures of the inside surfaces will not be very different from the inside air temperature and it would be reasonable to assume that t_{ai} is equal to t_{ei}. The total heat loss rate can then be found using the air temperatures.

Problem 15 *Fig 4* illustrates the plan of a college laboratory, situated on the first floor of a three storey building. Calculate the total heat loss rate assuming the following data:
height of room = 3 m
U-value of external walls = 1.3 W/m²K
U-value of glazing   = 5.7 W/m²K
area of glazing   = 30 m²
ventilation rate   = 1.5 air changes per hour
outside air temperature   = −1°C
inside air temperature   = 20°C
It may be assumed that (i) the temperatures of the rooms above and below the laboratory are the same as the laboratory temperature (ii) the air temperatures and environmental temperatures are the same.

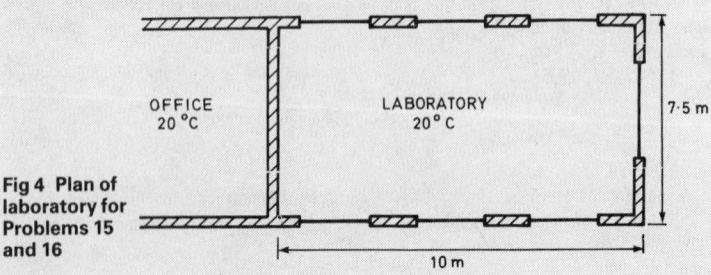

Fig 4 Plan of laboratory for Problems 15 and 16

Since the air temperatures are assumed to be the same as the environmental temperatures these can be used for finding the heat loss rates in the usual way:

Element	Area (m^2)	Heat loss rate $Q_u = AU(t_{ei} - t_{eo})$
External walls	52.5	$52.5 \times 1.3 \times (20 - (-1)) = 1433$
Glazing	30	$30 \times 5.7 \times (20 - (-1)) = 3591$
		Total $= 5024$ W

The heat loss rate by ventilation, assuming that the air and environmental temperatures are the same, is:

$$Q_v = \frac{1}{3} NV (t_{ai} - t_{ao})$$

$$= \frac{1}{3} \times 1.5 \times 225 \times (20 - (-1)) = 2363 \text{ W}$$

The total heat loss rate $= 5024 + 2363 = \mathbf{7387}$ **W**.

In the above calculation it was assumed that the air and environmental temperatures were the same. In practice this may not be true and errors of estimation of the heat loss rate could occur. The relationship between the inside air temperature and the inside environmental temperature depends on the type of heating system and the temperature selected as the design criterion. If the dry resultant temperature, t_c, is selected as the design criterion the total heat loss rate, Q_p, may be written as:

$$Q_p = (F_1 \Sigma(AU) + \frac{1}{3} F_2 NV)(t_c - t_{ao})$$

where $F_1 = \dfrac{t_{ei} - t_{ao}}{t_c - t_{ao}}$ and $F_2 = \dfrac{t_{ai} - t_{ao}}{t_c - t_{ao}}$

These equations allow the calculation of the heat loss rate and also the values of t_{ei} and t_{ai}. The values of F_1 and F_2 depend on the type of heating system and values for two types of heating systems are given in *Tables 2 and 3* below. More detailed tables are given in the CIBS Guide (Section A9–4). The following problem will illustrate the use of these tables.

> **Problem 16** For the laboratory described in *Problem 15* calculate the heat loss rate for a forced warm air heating system using the following temperatures:
> inside dry resultant temperature = 20°C
> outside air temperature = –1°C
> The rooms above and below the laboratory and also the office may be assumed to have a dry resultant temperature of 20°C.
> Determine also the inside environmental and inside air temperatures.

In order to find the values of F_1 and F_2 from *Table 2* it is necessary to find
(i) $\Sigma(AU)/\Sigma(A)$ for the external surfaces though which heat losses are occurring and
(ii) $NV/3\Sigma(A)$ where, in each case, $\Sigma(A)$ is the total surface area of the room. The calculation of $\Sigma(AU)$ can be as in *Table 4*.
The total surface area of the room, $\Sigma(A)$

$= 2(10 \times 3 + 7.5 \times 3 + 10 \times 7.5) = 255 \text{ m}^2$

Hence, $\Sigma(AU)/\Sigma(A) = 239.25/255 = 0.94$

TABLE 2 Value of F_1 and F_2 for 100% convective, 0% radiant (forced warm air heaters)

$\frac{NV}{3\Sigma(A)}$	$\Sigma(AU)/\Sigma(A)$											
	0.1		0.2		0.4		0.6		0.8		1.0		1.5	
	F_1	F_2	F_1	F_2	F_1	F_2	F_1	F_2	F_1	F_2	F_1	F_2	F_1	F_2
0.1	0.99	1.02	0.99	1.03	0.98	1.07	0.97	1.10	0.96	1.13	0.95	1.16	0.92	1.23
0.2	0.99	1.02	0.99	1.03	0.98	1.07	0.97	1.10	0.96	1.13	0.95	1.16	0.92	1.23
0.4	0.99	1.02	0.99	1.03	0.98	1.07	0.97	1.10	0.96	1.13	0.95	1.16	0.92	1.23
0.6	0.99	1.02	0.99	1.03	0.98	1.07	0.97	1.10	0.96	1.13	0.95	1.16	0.92	1.23
0.8	0.99	1.02	0.99	1.03	0.98	1.07	0.97	1.10	0.96	1.13	0.95	1.16	0.92	1.23

TABLE 3 Values of F_1 and F_2 for 50% convective, 50% radiant (single column radiators, floor warming systems, block storage heaters).

$\frac{NV}{3\Sigma(A)}$	$\Sigma(AU)/\Sigma(A)$											
	0.1		0.2		0.4		0.6		0.8		1.0		1.5	
	F_1	F_2	F_1	F_2	F_1	F_2	F_1	F_2	F_1	F_2	F_1	F_2	F_1	F_2
0.1	1.00	0.99	1.00	1.00	1.00	1.00	1.00	1.01	0.99	1.02	0.99	1.03	0.98	1.05
0.2	1.01	0.98	1.01	0.98	1.00	0.99	1.00	1.00	1.00	1.01	0.99	1.02	0.99	1.04
0.4	1.01	0.96	1.01	0.96	1.01	0.97	1.01	0.98	1.01	0.98	1.00	0.99	1.00	1.01
0.6	1.02	0.93	1.02	0.94	1.02	0.95	1.02	0.95	1.01	0.96	1.01	0.97	1.00	0.99
0.8	1.03	0.91	1.03	0.92	1.03	0.92	1.02	0.93	1.02	0.94	1.02	0.95	1.01	0.97

(Tables 2 and 3 are reproduced by the permission of the Chartered Institution of Building Services)

TABLE 4

Element	Area A	U-value	AU
External walls	52.5	1.3	68.25
Glazing	30	5.7	171
		$\Sigma(AU)$ =	239.25

In finding the value of $NV/3\Sigma(A)$ note that the volume of the room is 225 m³, thus:

$$\frac{NV}{3\Sigma(A)} = \frac{1.5 \times 225}{3 \times 255} = 0.44$$

From *Table 2* it will be found by interpolation that:
    $F_1 = 0.953$ and $F_2 = 1.15$
The heat loss rate is:

$$Q_p = (F_1 \Sigma(AU) + \frac{1}{3}F_2 NV)(t_c - t_{ao})$$
$$= (0.953 \times 239.25 + \frac{1}{3} \times 1.15 \times 1.5 \times 225)(20 - (-1))$$
$$= 4788 + 2717 = \textbf{7505 W}.$$

The inside environmental temperature can be found using:

$$F_1 = \frac{t_{ei} - t_{ao}}{t_c - t_{ao}}$$
$$0.953 = \frac{t_{ei} - (-1)}{20 - (-1)}$$
$$t_{ei} + 1 = 0.953 \times 21 = 20.01$$
$$t_{ei} = \textbf{19.01°C}$$

Similarly the inside air temperature may be found:

$$F_2 = \frac{t_{ai} - t_{ao}}{t_c - t_{ao}}$$
$$1.15 = \frac{t_{ai} - (-1)}{20 - (-1)}$$
$$t_{ai} = \textbf{23.15°C}$$

It is to be noted that the temperatures decreases in the order $t_{ai}, t_c, t_{ei}, t_{ao}$.

Problem 17 Repeat *Problem 16* where the laboratory is to be heated by a floor warming system.

From *Problem 16:*
    $\Sigma(AU)/\Sigma(A) = 0.94$ and $NV/3\Sigma(A) = 0.44$
Thus from *Table 3* it will be found by interpolation that:
    $F_1 = 1.004$ and $F_2 = 0.983$
The heat loss rate is:

$$Q_p = (F_1 \Sigma(AU) + \frac{1}{3}F_2 NV)(t_c - t_{ao})$$
$$= (1.004 \times 239.25 + \frac{1}{3} \times 0.983 \times 1.5 \times 225)(20 - (-1))$$
$$= 5044 + 2322 = \textbf{7366 W}$$

The inside environmental temperature is:

$$F_1 = \frac{t_{ei} - t_{ao}}{t_c - t_{ao}}$$

$$1.004 = \frac{t_{ei} - (-1)}{20 - (-1)}$$

$$t_{ei} = 20.08°C$$

The inside air temperature is:

$$F_2 = \frac{t_{ai} - t_{ao}}{t_c - t_{ao}}$$

$$0.983 = \frac{t_{ai} - (-1)}{20 - (-1)}$$

$$t_{ai} = 19.64°C$$

Note that, in this case, the temperatures decrease in the order t_{ei}, t_c, t_{ai}, t_{ao}. In all cases the values of t_{ei}, t_c and t_{ai} apply at the centre of the room, local variations may well occur as for example near a cold surface such as a window.

COMPOSITE CONSTRUCTIONS AND HEAT BRIDGES

Suppose that an element of a construction, for example a perimeter wall, is comprised of a number of different constructions having different areas and U-values as follows:
construction 1 of area A_1 and U-value U_1
construction 2 of area A_2 and U-value U_2
Then the average U-value denoted by $\bar{U}$ is given by:

$$\bar{U} = \frac{A_1 U_1 + A_2 U_2}{A_1 + A_2}$$

This formula is readily extended to any number of different constructions.

> *Problem 18* The external walls of a detached bungalow consist of 95 m² of cavity wall (U = 0.96 W/m² K) and 25 m² of single glazed windows (U = 5.7 W/m² K). Determine the average U-value of the perimeter walls.

Using the above formula:

$$\bar{U} = \frac{95 \times 0.96 + 25 \times 5.7}{95 + 25} = 1.95 \text{ W/m}^2\text{K}$$

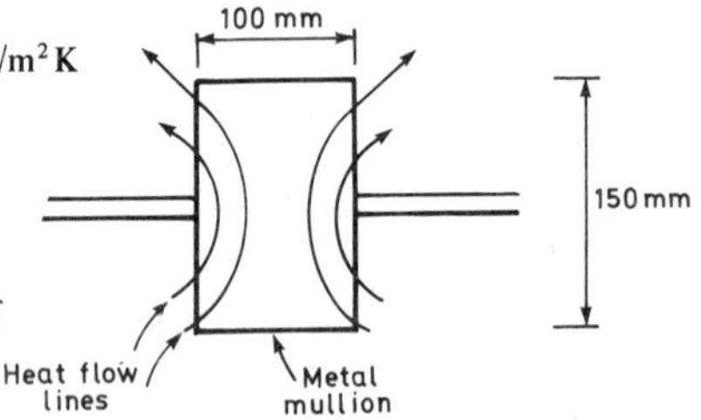

The above method can be used when an element of construction is bridged throughout its thickness by a material of different thermal resistance but it is not applicable when either the inner leaf or outer leaf only is bridged. The above method is also not directly applicable when there are projecting metal members, as illustrated in *Fig 5*. In this case it might be thought that the thermal

Fig 5 Heat flow through a metal mullion

resistance of the mullion could not be less than the sum of the inside and outside surface resistances which would suggest a U-value of 5.6 W/m² K. However the U-value of the mullion could be as high as 14 W/m² K depending on the projection inside and outside. This arises since the heat flow is not parallel to the edges.

Thus it can be seen that the bridging of a construction by materials of high conductivity, particularly metals, can lead to a serious increase in the U-value. In most cases the problem can be reduced by careful design and insulation of regions of low thermal resistance such as those that can occur in curtain wall construction at the joints of panels and the junction of panels with steel members.

REGULATIONS RELATING TO THERMAL INSULATION

The most important regulations are contained in parts F and FF of the Building Regulations. Part F relates to the thermal insulation of domestic dwellings and part FF to the conservation of fuel and power in buildings other than dwellings. The following does not purport to be a complete analysis of these regulations but only to illustrate the ideas contained in them.

Part F. Thermal insulation of dwellings

The basic essentials are as follows:
(i) The U-value of any part of a wall, floor or roof which encloses a dwelling, excluding any openings shall not exceed the value in the *Table 5*.
(ii) The calculated average U-value of perimeter wall, including any opening, shall not exceed 1.8 W/m² K.

TABLE 5 Maximum U-value of walls, floors and roofs

	Element of building	*Maximum U value of any part of element (in $W/m^2 K$)*
1	External wall	1.0
2	Wall between a dwelling and a ventilated space	1.0
3	Wall between a dwelling and a partially ventilated space	1.7
4	Wall between a dwelling and any part of an adjoining building to which Part F is not applicable	1.7
5	Wall or partition between a room and a roof space, including that space and the roof over that space	1.0
6	External wall adjacent to a roof space over a dwelling, including that space and any ceiling below that space	1.0
7	Floor between a dwelling and the external air	1.0
8	Floor between a dwelling and a ventilated space	1.0
9	Roof, including any ceiling to the roof or any roof space and any ceiling below that space	0.6

(Reproduced from the Building Regulations by the permission of the Controller, HMSO)

(iii) For calculating the average U-value of perimeter walling
(a) the U-value of any wall between a dwelling and another dwelling or between a dwelling and an internal space which is within the same building and not ventilated by permanent vents shall be assumed to be 0.5 W/m²K.
(b) The U-value for single glazing shall be assumed to be 5.7 W/m²K and 2.8 W/m²K for double glazing.

For full details and definitions the reader must consult the Building Regulations. The following problems will illustrate some of the calculations required.

> *Problem 19* A semidetached bungalow has internal dimensions 12 m by 10 m and a ceiling height of 2.5 m. The bungalow is attached to its neighbour on the 12 m side. The external cavity wall has a U-value of 0.9 W/m²K. The external wall includes 20 m² of single glazed windows. Determine whether this complies with thermal requirements for perimeter walls.

Note that areas of walling are considered to be inner areas. The following areas and U-value are appropriate:
external cavity wall : area = 60 m²; U-value = 0.9 W/m²K
single glazed window: area = 20 m²; U-value = 5.7 W/m²K
party wall : area = 30 m²; U-value = 0.5 W/m²K
The reader is advised to check the areas and to note that the U-values for glazing and party walls are given in the regulations. Using the previously given formula, the average U-value of the perimeter walls is:

$$\bar{U} = \frac{A_1 U_1 + A_2 U_2 + \ldots}{A_1 + A_2 + \ldots}$$

$$= \frac{60 \times 0.9 + 20 \times 5.7 + 30 \times 0.5}{60 + 20 + 30}$$

$$= 1.66 \text{ W/m}^2\text{K}$$

The perimeter walling complies with the requirements since the external cavity walling has a U-value of less than 1.0 W/m²K and the average U-value is less than 1.8 W/m²K.

> *Problem 20* The internal dimensions of a semidetached house are 10 m × 7 m on plan. The house is attached to its neighbour on the 10 m side. The external cavity wall has a U-value of 0.9 W/m²K. Determine the maximum permissible area of single glazing assuming that the external walls have a height of 5.2 m.

The relevant areas are:
Let area of the single glazing = A
Area of cavity wall = 5.2 (10 + 7 + 7) − A = 124.8 − A
Area of party wall = 52 m².
Total area of perimeter wall = 176.8 m².
The U-values of the glazing and party walls are to be taken as 5.7 and 0.5 W/m²K respectively. Using the formula for the average U-value:

$$\bar{U} = \frac{(124.8 - A) \times 0.9 + 52 \times 0.5 + A \times 5.7}{176.8}$$

$$= \frac{112.3 - 0.9A + 26 + 5.7A}{176.8} = \frac{138.3 + 4.8A}{176.8}$$

It is required that the average U-value must not exceed $1.8 \text{ W/m}^2\text{K}$. Hence:

$$\frac{138.3 + 4.8A}{176.8} \leq 1.8$$

$$138.3 + 4.8A \leq 1.8 \times 176.8$$
$$138.3 + 4.8A \leq 318.24$$
$$4.8A \leq 179.94$$
$$A \leq 37.49 \text{ m}^2$$

Problem 21 A semi-detached house has external cavity walls with a U-value of $0.9 \text{ W/m}^2\text{K}$. The party wall constitutes 30% of the area of the perimeter walls. The wall between the house and the attached garage, which constitutes 20% of the area of the perimeter walls, may be taken to have a U-value of $1.7 \text{ W/m}^2\text{K}$. Determine the percentage of the external wall that can be single glazing.

Let the percentage of single glazing be $x\%$, then the percentage of external cavity wall is $(50 - x)$ since the cavity wall and windows together form 50% of the perimeter wall. Using the percentage areas, the average U-value is:

$$\bar{U} = \frac{30 \times 0.5 + 20 \times 1.7 + 0.8(50 - x) + 5.7x}{100}$$

$$= \frac{15 + 34 + 40 - 0.8x + 5.7x}{100}$$

$$= \frac{89 + 4.9x}{100}$$

Note that the U-values for the party wall and the single glazing are 0.5 and $5.7 \text{ W/m}^2\text{K}$ as specified in the Regulations. If the average U-value is not to exceed $1.8 \text{ W/m}^2\text{K}$ then:

$$\frac{89 + 4.9x}{100} \leq 1.8$$
$$89 + 4.9x \leq 180$$
$$4.9x \leq 91$$
$$x \leq 18.5\%$$

Thus the area of single glazing must be less than 18.5% of the perimeter walls. If this is expressed as a percentage of the external wall, excluding the party wall and the garage wall, the percentage of single glazing is:

$$\frac{18.5}{100} \times \frac{100}{(100 - 50)} = 37\%$$

Thus 37% of the external wall can be single glazing. The reader is now advised to read Schedule 11 of the Building Regulations.

Part FF

For the purpose of the Building Regulations buildings are divided into eight Purpose Groups, these are:

Group	Building Type
I	Small residential (private houses)
II	Institutional (hospitals etc.)
III	Other residential (excluding groups I and II)
IV	Office
V	Shop

VI	Factory
VII	Other place of assembly (educational, recreational)
VIII	Storage and general.

Buildings in purpose group I are covered by Part F of the Regulations and are thus exempt from Part FF. The reader is advised to consult the Regulations for the other exemptions, for example the exemptions for some Group III buildings, small buildings of floor area less than 30 m², and buildings for which the heating system has only a small output.

The Regulation requires that the enclosing structure provides adequate resistance to the flow of heat. The deemed to satisfy provisions provide a conventional approach to the provision of an adequate resistance to the flow of heat. This is achieved in two ways, firstly by limiting the single glazed areas, expressed as percentages of the wall and roof areas, to the following maximum values:

(Note that wall and roof areas are calculated from internal, not external, dimensions)

Group	*II, III*	*IV, V, VII*	*VI, VIII*
Window opening	25	35	15
Rooflight opening	20	20	20

Higher percentages are acceptable if the calculated heat loss through all openings does not exceed that which would occur if the openings had been single glazed and the values in the above table complied with. Thus double glazing will permit a considerable increase in the allowable percentage openings.

The second method is to limit the U-values of walls, roofs and floors with the underside exposed to the air to 0.6 W/m²K except that 0.7 W/m²K is permitted for group VI or VIII (if used for storage). Higher U-values of parts of the construction are allowed as long as the calculated rate of heat loss through all such walls, floors and roofs does not exceed that which would occur had the U-values complied with the above figures.

The following U-values are assumed for window or rooflight openings:

5.7 W/m²K	if single glazed
2.8 W/m²K	if double glazed
2.0 W/m²K	if triple glazed

Problem 22 A small factory has roof and wall areas, inclusive of openings, of 1000 m² and 4000 m² respectively. If 10% of the roof is to be single glazed roof lights determine:
(i) the area of single glazing permissible in the walls;
(ii) the area of double glazing permissible in the walls;
(iii) the area of single glazing permissible in the walls if one elevation of the building is to contain 800 m² of double glazing.

(i) The area of rooflights = $\dfrac{10}{100} \times 1000 = 100$ m²

The areas permitted for a purpose group VI building are 15% for windows and 20% for rooflight openings giving the following values:

permissible area of rooflights = $\dfrac{20}{100} \times 1000 = 200$ m²

permissible area of windows = $\dfrac{15}{100} \times 4000 = 600$ m²

As these are both to be single glazed the total area is 800 m², however **100 m² of** rooflights are to be used which leaves **700 m²** for window openings in the external wall.

(ii) In this case, since double glazing is to be used, the U-values must be taken into account. From above, 700 m² of single glazing can be used. Thus, remembering that the U-values for single and double glazing are 5.7 and 2.8 W/m²K, the heat loss will be the same if:

2.8 × area of double glazing = 5.7 × area of single glazing
2.8 × area of double glazing = 5.7 × 700

area of double glazing $= \dfrac{5.7}{2.8} \times 700 = \mathbf{1425\ m^2}$

(iii) Let A = area of single glazing and extending the principles in (ii) above the heat loss will be the same if:

800 × 2.8 + 5.7 × A = 700 × 5.7
5.7A = 700 × 5.7 − 800 × 2.8
5.7A = 1750
A = **307 m²**

Problem 23 A small factory has roof and wall areas, inclusive of openings of 1000 m² and 4000 m² respectively. 15% of the roof is to be single glazed rooflights. 600 m² of the external walls are to be double glazed windows. The remainder of the maximum permissible areas of rooflights and windows is to be single glazed windows. If the U-value of the roof construction is 1.1 W/m²K determine the necessary U-value of the external wall construction.

It is necessary to detail the areas of rooflights and glazing in order to find the areas of the roof and walls.

For a purpose group VI building:

maximum permissible area of rooflights = $\dfrac{20}{100} \times 1000 = 200\ m^2$

maximum permissible area of windows = $\dfrac{15}{100} \times 4000 = 600\ m^2$

area of rooflights used = $\dfrac{15}{100} \times 1000 = 150\ m^2$

Using the specified U-values and ensuring that heat loss will be the same as if single glazing were used to the maximum permitted:

150 × 5.7 + 600 × 2.8 + 5.7A = 800 × 5.7

where A = area of single glazing

5.7A = 2025
A = 335 m²

Hence:
area of roof = 1000 − 150 = 850 m²
area of walls = 4000 − 600 − 335 = 3065 m²

The U-value of the roof exceeds the value allowed for roofs and walls for a purpose group VI building which is 0.7 W/m²K. However this will be satisfactory as long as the heat loss through the walls and roof does not exceed that which would occur if the U-value were 0.7 W/m²K, hence:

850 × 1.1 + 3065U = (850 + 3065) × 0.7

where U is the required U-value of the wall.

3065U = 1805.5
U = **0.59 W/m²K**

HEATING FUELS

If cost comparisons are to be made it is essential to understand the methods of pricing commonly employed.

(i) **Electricity.** The unit is the kilowatt hour.
$$1 \text{ kWh} = 1 \text{ kW} \times 1 \text{ hour}$$
$$= 1000 \text{ W} \times 3600 \text{ s} = 3.6 \text{ MJ}$$

(ii) **Gas.** Gas is sold by the therm; 1 therm being 105.5 MJ. Most gas meters are calibrated in hundreds of cubic feet. The numbers of therms used can be calculated from:

$$\text{number of therms} = \frac{\text{cubic feet of gas} \times \text{calorific value}}{100000}$$

The calorific value being quoted in Btu/ft^3 and having a typical value of 1046 for natural gas, which corresponds to 39 MJ/m^3.

(iii) **Oil.** Oil is sold by the litre, several grades of heating oil are available depending on the type of boiler used. In order to determine the amount of heat produced it is necessary to know both the calorific value, in MJ/kg, and the relative density of the oil. If these are known the heat produced by 1 litre is found as:

heat produced = relative density × calorific value MJ/litre

(iv) **Solid fuel.** Solid fuel is sold by the tonne, the calorific value depending on the grade of coal. A value of 27 MJ/kg would be appropriate for general purpose coal.

Problem 24 For the bungalow in *Problem 13* use the result of *Problem 13(a)* and *Problem 14* to calculate:
(a) the total heat loss in a 24-hour period.
(b) the number of therms of gas used by a boiler having an efficiency of 75%.
(c) the number of litres of oil, of relative density 0.79 and calorific value 43.4 MJ/kg, used by a boiler of efficiency 75%.
(d) the number of kilograms of coal, of calorific value 27 MJ/kg, used by a boiler having an efficiency of 60%.

In *Problem 13(a)* the heat loss rate through the construction was found to be 8834 W. In *Problem 14* the heat loss rate due to ventilation was found to be 2851 W. Thus the total heat loss rate is 8834 + 2851 = 11685 W.

(a) Heat loss = heat loss rate × time in seconds
$$= 11685 \times 24 \times 3600$$
$$= 1009584000 \text{ J} = \mathbf{1009.6 \text{ MJ}}$$

(b) To find the number of therms of gas it must be remembered that 1 therm = 105.5 MJ.
$$\text{Number of therms} = \frac{1009.6}{105.5} = 9.57$$

This assumes that the boiler is 100% efficient and as it is not, more heat will need to be provided:

$$\text{Number of therms} = 9.57 \times \frac{100}{75} = \mathbf{12.76 \text{ therms}}$$

(c) Heat produced by 1 litre of oil = 0.79 × 43.4 = 34.3 MJ.
$$\text{Number of litres} = \frac{1009.6}{34.3} \times \frac{100}{75} = \mathbf{39.2 \text{ litres}}$$

Note that the boiler efficiency has been taken into account.

(d) Quantity of coal = $\frac{1009.6}{27} \times \frac{100}{60} = \mathbf{62.3 \text{ kg}}$

TEMPERATURE DISTRIBUTION

In order to study condensation problems it is necessary to determine the temperatures throughout the construction. For the simple construction shown in *Fig 6* the inside, outside and interface temperatures are denoted by t_i, t_1, t_2 and t_o. The thermal resistances are R_{si}, R_1 and R_{so}. If R denotes the total thermal resistance then:

$$\frac{t_i - t_o}{R} = \frac{t_i - t_1}{R_{si}} = \frac{t_1 - t_2}{R_1} = \frac{t_2 - t_o}{R_{so}}$$

In practice it would be necessary to find t_1 and t_2 when all other quantities are known. Hence:

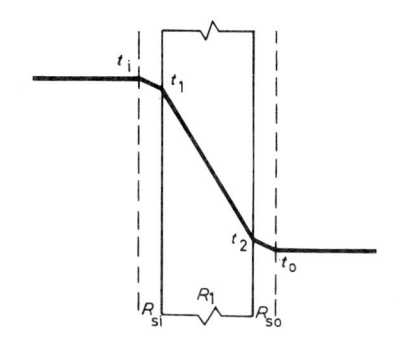

Fig 6 Temperature distribution in a structure

$$\frac{t_i - t_o}{R} = \frac{t_i - t_1}{R_{si}}$$

which gives on rearrangement:

$$t_i - t_1 = \frac{R_{si}}{R}(t_i - t_o)$$

Similarly,

$$t_1 - t_2 = \frac{R_1}{R}(t_i - t_o)$$

It will be seen that temperature drop across each element of the construction is given by:

$$\text{temperature drop} = \frac{\text{resistance of element}}{\text{total resistance}} \times \text{total temperature drop}$$

The following problem should clarify the method.

> *Problem 25* A solid brick wall has the following properties:
> thermal resistance of brickwork = 0.22 m² K/W
> R_{si} = 0.12 m² K/W
> R_{so} = 0.05 m² K/W
> If the inside and outside temperatures are 22°C and 4°C respectively find the surface temperatures of the brickwork and illustrate the temperature distribution graphically.

The calculations are set out in *Table 6*.
The following points should be noted:
(i) the total temperature drop = 22 − 4 = 18°C
(ii) each temperature drop has been calculated by:

$$\text{temperature drop} = \frac{\text{resistance of element}}{\text{total resistance}} \times \text{total temperature drop}$$

The total resistance being shown at the foot of the column of resistances.
(iii) The actual temperature is found by successively subtracting the temperature drops from the inside temperature of 22°C. The final temperature being the outside temperature.

TABLE 6

Element	Thermal resistance $(m^2 k/W)$	Temperature drop across element (K)	Temperature $(°C)$
R_{si}	0.12	$\frac{0.12}{0.39} \times 18 = 5.54$	$22 - 5.54 = 16.46$
Brickwork	0.22	$\frac{0.22}{0.39} \times 18 = 10.15$	$16.46 - 10.15 = 6.31$
R_{so}	0.05	$\frac{0.05}{0.39} \times 18 = 2.31$	$6.31 - 2.31 = 4.00$
Totals	0.39	18.00	

(iv) The sum of the temperature drops is equal to the total temperature drop of 18°C. This provides a check on the accuracy of calculation.
(v) The calculation must take the elements of the construction in order.

The temperature distribution is shown in *Fig 7*.

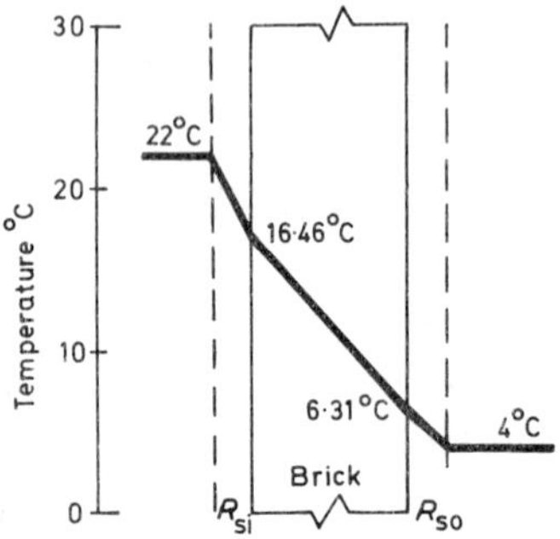

Fig 7 Temperature distribution for Problem 25

Problem 26 Details of the construction of a cavity wall are given in *Table 7*. Determine the temperature distribution throughout the construction assuming that the inside and outside temperatures are 19°C and 2°C respectively.

TABLE 7

Element	Thickness (m)	Thermal conductivity λ (W/mK)	Thermal resistance $(m^2 K/W)$
R_{si}	—	—	0.123
Plaster	0.013	0.4	—
Aerated blocks	0.100	0.17	—
Cavity (R_a)	—	—	0.18
Brickwork	0.102	0.9	—
R_{so}	—	—	0.055

As before the calculations are best set out in tabular form, see *Table 8*.
The following points should be noted:
(i) the thermal resistance of each solid element of the construction is calculated as $R = l/\lambda$.

34

TABLE 8

Element	Thermal resistance $(m^2 k/W)$	Temperature drop across element (K)	Temperature $(°C)$
R_{si}	= 0.123	$\frac{0.123}{1.092} \times 17 = 1.91$	$19 - 1.91 = 17.09$
Plaster	$0.013/0.4 = 0.033$	$\frac{0.033}{1.092} \times 17 = 0.51$	$17.09 - 0.51 = 16.58$
Blocks	$0.1/0.17 = 0.588$	$\frac{0.588}{1.092} \times 17 = 9.15$	$16.58 - 9.15 = 7.43$
Cavity	= 0.18	$\frac{0.18}{1.092} \times 17 = 2.80$	$7.43 - 2.80 = 4.63$
Brickwork	$0.102/0.9 = 0.113$	$\frac{0.113}{1.092} \times 17 = 1.76$	$4.63 - 1.76 = 2.87$
R_{so}	= 0.055	$\frac{0.055}{1.092} \times 17 = 0.86$	$2.87 - 0.86 = 2.01$
Totals	*1.092*	*16.99*	

(ii) the total temperature drop should be 17 K. The 16.99 K obtained is consistent with an accuracy of two decimal places. The rounding off errors also give a final temperature of 2.01°C and not 2°C. The temperature distribution is shown in *Fig 8*.

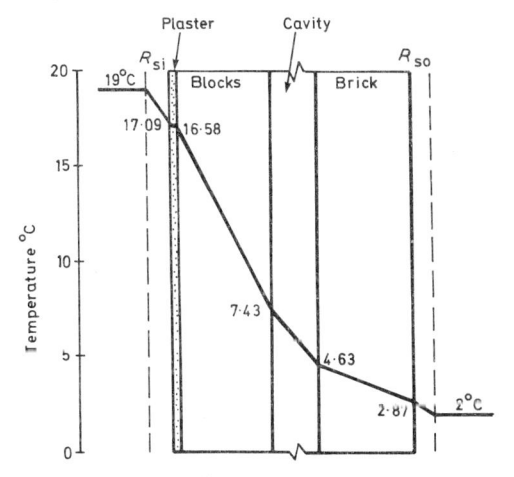

Fig 8 Temperature distribution for Problem 26

PATTERN STAINING

This is caused by the existence of different surface temperatures due to varying thermal properties of the structure. The consequence of these temperature variations may be twofold:

(a) the regions of low surface temperature are more likely to be below the dew point temperature of the air so that condensation may occur.
(b) dust particles will settle preferentially on the cooler parts of the surface since the molecules possess lower thermal energy causing regions of light and dark, termed pattern staining.

To obviate both these effects it is advisable to ensure that the thermal resistance of the construction is as uniform as possible.

> *Problem 27* Fig 9 illustrates a construction having the following properties:
> Material A has a thermal conductivity of 0.1 W/mK
> Material B has a thermal conductivity of 1.2 W/mK
>
> **Fig 9 Construction for Problem 27**
>
> Assuming the inside and outside surface resistances are 0.12 and 0.06 m²K/W respectively and that the inside and outside temperatures are 20°C and 5°C respectively estimate the surface temperature of the construction at (i) X; (ii) Y. Comment on the likelihood of pattern staining.

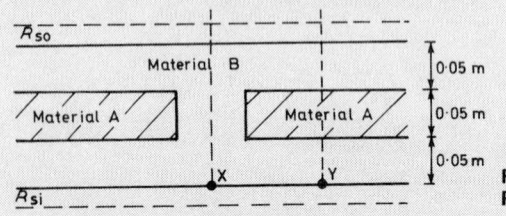

Consider the section at X:

Total resistance = $0.12 + \frac{0.15}{1.2} + 0.06 = 0.305$ m²K/W

Temperature drop across inside surface resistance

$= \frac{0.12}{0.305} \times (20 - 5)$

$= 5.9$ K

Surface temperature at X = 20 − 5.9 = 14.1°C

Consider the section at Y:

Total resistance = $0.12 + \frac{0.05}{1.2} + \frac{0.05}{0.1} + \frac{0.05}{1.2} + 0.06$

$= 0.763$ m²K/W

Temperature drop across inside surface resistance = $\frac{0.12}{0.763} \times (20 - 5)$

$= 2.36$ K

Surface temperature at Y = 20 − 2.36 = 17.64°C

It will be noted that the surface temperature at Y is about 3.5°C higher than that at X and such a difference is likely to cause pattern staining.

EXERCISES (answers on page 184)

1. The thermal conductivity of a material is 0.8 W/mK when its moisture content is 2%. Estimate the thermal conductivity of the material when (a) the moisture is 5%; (b) the material is dry.

2 In the measurement of the thermal conductivity of a board material a sample of area of 0.04 m² and thickness 25 mm was used. When the heat flow rate was 2 W the temperature difference between the faces was 6.8 K. Calculate the thermal conductivity of the material.

3 The thermal resistance of 0.11 m thickness of brickwork is 0.18 m²K/W when it is dry. Calculate the thermal conductivity of the brickwork. Estimate its thermal resistance when the moisture content is 3%.

4 Calculate the internal surface resistance in the following cases:
   (i) for a floor with heat flow downwards, the convection conductance being 1.5 W/m²K.
   (ii) for a ceiling with heat flow upwards, the convection conductance being 4.3 W/m²K.
   In each case assume the radiation conductance to be 5.7 W/m²K and the emissivity factor to be 0.9.

5 In the following select the correct option: the internal surface resistance for surfaces with the same emissivity factor is:
   (a) the same for all surfaces;
   (b) is less for a wall with heat flow horizontal than for a ceiling with heat flow upwards;
   (c) is less for a floor with heat flow downwards than for a wall with heat flow horizontal;
   (d) is less for a ceiling with heat flow upwards than for a floor with heat flow downwards.

6 In the following select the correct options: the external surface resistance:
   (a) does not depend on windspeed;
   (b) will depend on windspeed;
   (c) will be higher for surfaces of buildings on exposed hill tops than for similar buildings in sheltered urban areas;
   (d) will be lower for materials of high emissivity than for materials of low emissivity in the same situation.

7 Calculate the U-value for an external wall of single leaf construction consisting of:
   200 mm lightweight concrete block (λ = 0.19 W/mK)
   20 mm glass fibre quilt (λ = 0.04 W/mK)
   10 mm plasterboard (λ = 0.16 W/mK)
   Assume the inside and outside surface resistances are 0.12 m²K/W and 0.06 m²K/W respectively.

8 Calculate the U-value of an external cavity wall consisting of:
   105 mm brickwork (λ = 0.84 W/mK)
   50 mm cavity of thermal resistance 0.18 m²K/W
   100 mm lightweight block (λ = 0.19 W/mK)
   13 mm lightweight plaster (λ = 0.16 W/mK)
   Assume the inside and outside surface resistances are 0.12 m²K/W and 0.06 m²K/W respectively.

9 For the construction in *Exercise 8* find the U-value if the cavity is filled with foam of thermal conductivity 0.04 W/mK.

10 An industrial roof consists of asbestos cement sheet, a loft space and a ceiling of

plasterboard and has a U-value of 2.4 W/m²K. Find the thickness of fibreglass quilt (λ = 0.04 W/mK) to be laid on the upperside of the ceiling in order to reduce the U-value to 0.6 W/mK. Assume that the thermal resistance of the loft space is unaltered by the fibreglass quilt.

Determine also the heat loss rate before and after insulation through 50 m² of this roof if the inside and outside environmental temperatures are 15°C and 2°C respectively.

11 A semi-detached house has the following properties (see *Table 9*):
The volume of the house is 211 m³ and the ventilation rate is 1.5 air changes per hour. Calculate the heat loss rate for inside and outside temperatures of 18°C and −1°C respectively; it may be assumed that the inside air and environmental temperatures are the same. Express each component of the heat loss rate as a percentage of the total heat loss rate.

TABLE 9

Element	Area (m²)	U-value (W/m²K)
External walls	71	1.5
Windows	20	5.3
Floor	45	0.65
Roof	45	1.7

12 A corner office in a multistorey block has two exposed walls which consist of 16 m² single glazing (U = 5.7 W/m²K) and 11 m² of wall construction (U = 1.0 W/m²K). The office is 5 m by 4 m on plan and is 3 m high. The ventilation rate is 1.5 air changes per hour. The room is to be heated to a dry resultant temperature of 20°C when the outside air temperature is −1°C. If heat losses to surrounding offices, including those above and below may be ignored calculate the heat loss rate for a forced warm air heating system. Determine also the inside environmental and air temperatures.

13 Find the average U-value for a construction which consists of 20 m² of cavity wall having a U-value of 0.8 W/m²K and 8 m² of double glazing have a U-value of 2.8 W/m²K.

14 A centre terraced house is 9 m by 6.5 m on plan and is attached on the 9 m sides. The height of the first floor ceiling above the ground floor is 4.9 m. The external walls are of a cavity construction (U = 0.90 W/m²K) and contain 40% single glazing (U = 5.7 W/m²K). The U-value of the party walls may be taken as 0.5 W/m²K.
Determine the average U-value of the perimeter walls.

15 The external walls of a detached building are of cavity construction having a U-value of 0.8 W/m²K. Determine the maximum percentage of single glazing (U = 5.7 W/m²K) that is permissible if the average U-value is not to exceed 1.8 W/m²K.

16 Repeat *Exercise 15* for double glazing (U = 2.8 W/m²K).

17 Repeat *Exercise 15* for the U-value of the cavity construction ranging from 0.5 to 1.0 W/m²K in steps of 0.1. Construct a suitable graph to show the variation of percentage single glazing against the U-value of the cavity wall.

18 The perimeter walls of a building are:
    Party wall: 25% of area of perimeter wall; U-value = 0.5 W/m²K.
    Wall between house and partly ventilated space: 15% of area of perimeter wall; U-value = 1.5 W/m²K
    External cavity wall: U-value 0.85 W/m²K
    Single glazing: U-value 5.7 W/m²K
    Determine the maximum percentage single glazing in the external cavity wall that is permissible if the average U-value of the perimeter walls is not to exceed 1.8 W/m²K.

19 Repeat *Exercise 18* for a double glazing having a U-value of 2.8 W/m²K.

20 An office building is 25 m by 15 m wide based on internal dimensions. It has four storeys with an internal floor to ceiling height of 2.8 m.
    (i)   Determine the maximum permissible area of single glazed window and rooflight openings.
    (ii) If no rooflights are used determine the maximum area of single glazing permissible in the walls.
    (iii) If no rooflights are used determine the maximum area of double glazing permissible in the walls.
    (iv) If 50 m² of single glazed rooflights are used and one long facade is completely double glazed find the maximum area of single glazing permissible in the remaining facades.
    (v)   In each of the cases (i) to (iv) above, assuming that the maximum permissible area of glazing is used, determine the necessary U-value of the wall construction if the U-value of the roof is 0.9 W/m²K. (Warning: beware case (iii))

21 The heat loss from a building is 7 kW on average over a 24 hour period. If the heating is provided by an oil fired boiler find the number of litres of oil used assuming that: the oil has a relative density of 0.79 and a calorific value of 43.5 MJ/kg and that the boiler is 70% efficient.

22 Calculate the temperature distribution for the wall in *Exercise 7* when the inside and outside temperatures are 19°C and 1°C respectively.

23 Calculate the temperature distribution for the wall in *Exercise 8* when the inside and outside temperatures are 19°C and –1°C respectively.

3 Condensation

SURFACE AND INTERSTITIAL CONDENSATION

Everybody is familiar with surface condensation on windows and other cold surfaces. Interstitial condensation, which occurs within the materials of the construction may be more serious. Interstitial condensation may cause degradation of the materials in the form of mould, rot and corrosion and will also cause a reduction in the insulation value of the construction.

Except when warm moist air rapidly follows a cool dry spell of weather condensation is due to the excess moisture produced by the occupants of the building and their activities. For example a person breathes out more than a litre of water, as water vapour, in a day. Washing, cooking and many industrial activities produce large quantities of water vapour. Natural ventilation will tend to remove excess water vapour but research shows that at normal ventilation rates the moisture excesses, expressed in kg/kg dry air, that can be expected in practice are:

office, shops, classrooms	0.0017 kg/kg dry air
dwellings	0.0034 kg/kg dry air
catering establishments	0.0068 kg/kg dry air

Prediction of surface condensation

Condensation will occur if the moist air is cooled below the dew point temperature. A possible prediction method is illustrated in the next problem.

> *Problem 1* A window in a house has a U-value of 5.6 W/m²K. The inside surface resistance $R_{si} = 0.12$ m² K/W. If the outside air is at 1°C and 90% relative humidity determine whether condensation will occur on the inside of the window when the inside temperature is 15°C.

It is necessary in the first instance to calculate the surface temperature of the window. The total resistance $R = 1/U$ as previously explained.

$$R = \frac{1}{U} = \frac{1}{5.6} = 0.179 \text{ m}^2 \text{ K/W}.$$

The temperature drop across the inside surface resistance is then given by:

$$\text{temperature drop} = \frac{0.12}{0.179} \times (15 - 1) = 9.4 \text{ K}$$

Thus the inside surface temperature of the window is $15 - 9.4 =$ **5.6°C**.

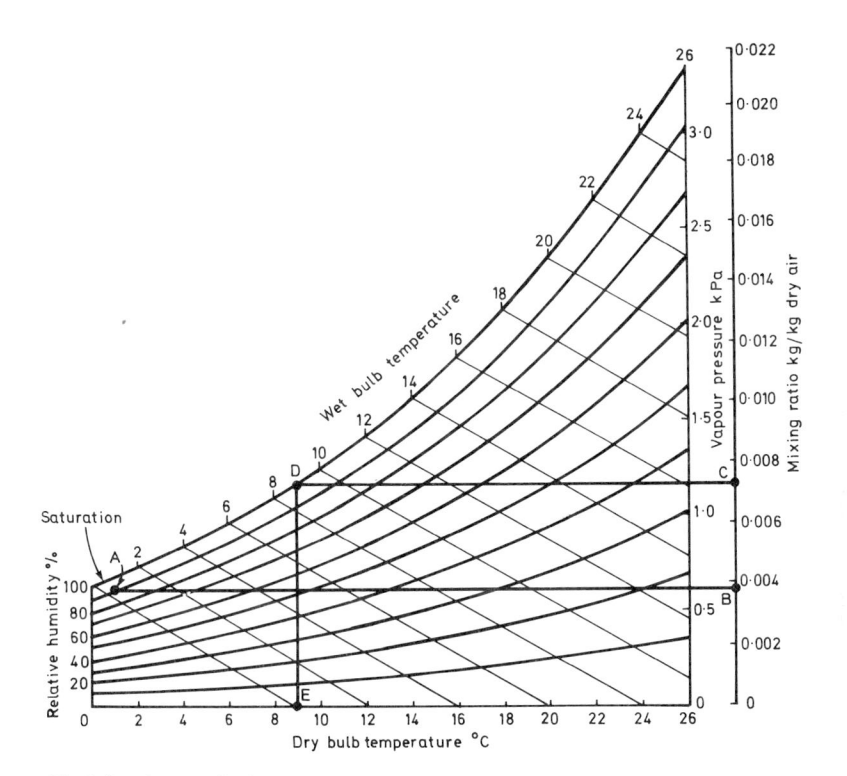

Fig 1 Psychrometric chart

It is now necessary to determine the outside and inside vapour conditions. *Fig 1* shows a simplified psychrometric chart which relates the air temperature, the relative humidity and water vapour pressure. There are two scales on the right hand side which allows conversion between vapour pressure and mixing ratio. The point A on *Fig 1* shows the condition of the outside air which is at 1°C and 90% relative humidity. Reading across to the right hand scales it will be seen that the air has a mixing ratio of 0.0037 kg/kg dry air with a corresponding vapour pressure of 0.6 kPa as shown by point B.

Since the window is in a house the moisture excess is 0.0034 kg/kg dry air. Thus:
  mixing ratio of inside air = 0.0037 + 0.0034 = 0.0071
This mixing ratio is shown by point C on *Fig 1*. In order to find the dew point temperature for the inside air a horizontal line is drawn at the mixing ratio of 0.0071 until it cuts the 100% relative humidity curve at D. The corresponding temperature at E, which is 9°C, is the dew point temperature. The temperature of the surface of the window is 5.6°C which is below the dew point temperature of 9°C thus condensation will occur on the window.

Prediction of interstitial condensation

It is necessary to:
(i) calculate the temperature distribution through the construction;
(ii) calculate the vapour pressure distribution through the construction and hence arrive at the dew point temperature distribution;
(iii) compare the temperature and dew point temperature distributions and thus estimate the likelihood of condensation.

Before proceeding, item (ii) above needs careful consideration. Many materials will permit the diffusion of water vapour through them, and when a vapour pressure difference exists across the material water vapour will flow through it. Materials vary widely in the resistance that they provide to the flow of vapour. The fundamental property of the material is the vapour permeability which is analogous to thermal conductivity for heat flow.

The vapour permeability is the quantity of water vapour, in kilogrammes, that passes in one second through one metre thickness of material having a cross sectional area of 1 m² when a vapour pressure difference of 1 Pa exists across the faces. The units of this quantity are kg m/m²s Pa, which are often simplified to kg m/Ns. The vapour permeability will be denoted by δ.

The vapour resistivity, denoted by r_v, which is often used, is the reciprocal of the vapour permeability. This is an analogous definition to that of thermal resistivity. The units of vapour resistivity are Ns/kg m.

TABLE 1

Material	Vapour resistivity r_v (Ns/kg m)
Brickwork	25–100 $\times 10^9$
Concrete	30–100 $\times 10^9$
Plaster	35–60 $\times 10^9$
Timber	45–75 $\times 10^9$
Hardboard	400–1000 $\times 10^9$
Foamed urea-formaldehyde	20–30 $\times 10^9$

Table 1 gives values of vapour resistivity for some materials. It will be observed, that for some materials, the figures vary widely depending on their structure.

The rate of vapour flow through a material is given by:

$$G = \frac{\delta A(p_i - p_o)}{l}$$

where

G = rate of vapour flow in kg/s;
δ = vapour permeability in kg m/Ns;
A = area in m²;
l = thickness in m;
p_i, p_o = inside and outside vapour pressures in Pa (N/m²).

The reader is requested to compare this formula with the formula for heat conduction:

$$Q = \frac{\lambda A(t_i - t_o)}{l}$$

Note the similarities between (i) thermal conductivity and vapour permeability and

(ii) temperature and vapour pressure. In an exactly analogous manner to the thermal case the vapour resistance, R_v, is defined as:

$$R_v = \frac{\text{thickness}}{\text{vapour permeability}} = \frac{l}{\delta} = l \times \left(\frac{1}{\delta}\right) = lr_v$$

This can be expressed in words as:

Vapour resistance = thickness × vapour resistivity

For composite constructions, the total vapour resistance is found by adding the individual vapour resistances. As the law governing the flow of vapour is analogous to the law governing the flow of heat the method of calculating the vapour pressure distribution is analogous to the method of calculating the temperature distribution. The method is illustrated in the next problem.

Problem 2 Determine the likelihood of interstitial condensation in a wall consisting of 100 mm brick ($\lambda = 0.84$ W/mK, $r_v = 150 \times 10^9$ Ns/kg m) lined internally with 12 mm insulating board ($\lambda = 0.06$ W/mK, $r_v = 9 \times 10^9$ Ns/kg m).

The inside and outside thermal resistances may be taken as 0.12 m²K/W and 0.06 m²K/W respectively. The outside air temperature is 2°C and the relative humidity is 90%. The inside air temperature is 22°C and a vapour excess of 0.004 kg/kg dry air may be assumed.

Stage (i) The temperature distribution can be calculated in the usual way as shown in *Table 2*. The temperature distribution is shown, along with the construction in *Fig 2*.
Stage (ii) It is necessary in the first instance to determine the inside and outside vapour pressures. The point A on *Fig 3* shows the condition of the outside air at 2°C and 90% relative humidity.

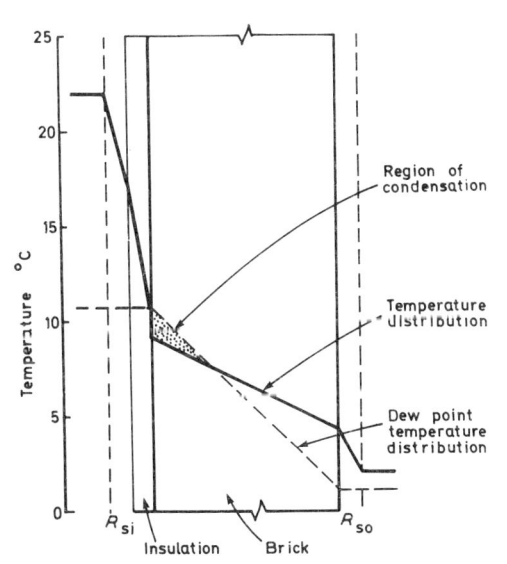

Fig 2 Temperature and dew point temperature distributions for Problem 2

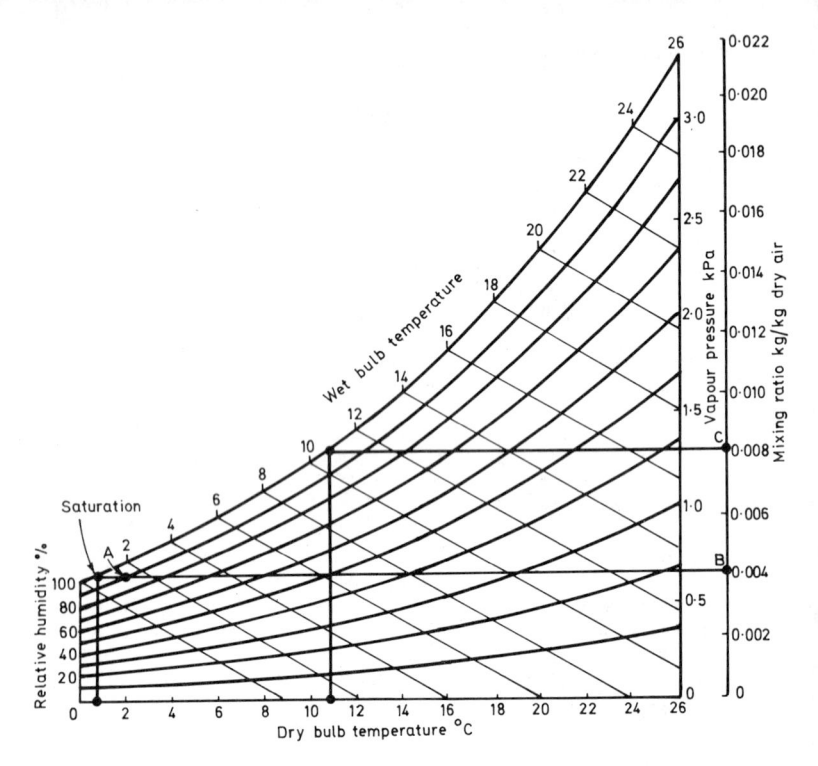

Fig 3 Psychrometric chart for Problem 2

TABLE 2

Element	Thermal resistance ($m^2 K/W$)	Temperature drop across element (K)	Temperature (°C)
R_{si}	0.12	$\frac{0.12}{0.5} \times 20 = 4.8$	$22 - 4.8 = 17.2$
Insulation	$\frac{0.012}{0.06} = 0.20$	$\frac{0.20}{0.5} \times 20 = 8.0$	$17.2 - 8.0 = 9.2$
Brick	$\frac{0.100}{0.84} = 0.12$	$\frac{0.12}{0.5} \times 20 = 4.8$	$9.2 - 4.8 = 4.4$
R_{so}	= 0.06	$\frac{0.06}{0.5} \times 20 = 2.4$	$4.4 - 2.4 = 2.0$
Totals	0.5	20.0	

TABLE 3

Element	Vapour resistance (Ns/kg)	Vapour pressure drop (kPa)	Vapour pressure (kPa)
Insulation	$0.012 \times 9 \times 10^9 = 0.1 \times 10^9$	$\dfrac{0.1 \times 10^9}{15.1 \times 10^9} \times 0.64 = 0.004$	$1.29 - 0.004 = 1.286$
Brick	$0.100 \times 150 \times 10^9 = 15 \times 10^9$	$\dfrac{15 \times 10^9}{15.1 \times 10^9} \times 0.64 = 0.636$	$1.286 - 0.636 = 0.65$
Totals	$= 15.1 \times 10^9$	0.64	

Notes: (i) The vapour resistances have been calculated using:
vapour resistance = thickness × vapour resistivity.

(ii) The inside and outside vapour pressures are 1.29 kPa and 0.65 kPa respectively, thus the vapour pressure drop across the construction is: $1.29 - 0.65 = 0.64$ kPa.

(iii) The vapour pressures are found by successively finding the vapour pressure drops from the inside vapour pressure of 1.29 kPa.

By reading across to the right hand scales it will be seen that the air has a mixing ratio of 0.004 kg/kg dry air, at point B, with a corresponding vapour pressure of 0.65 kPa. The mixing ratio of the inside air is $0.004 + 0.004 = 0.008$ kg/kg dry air, since a vapour excess of 0.004 kg/kg dry air has been specified. Using the right hand scales this corresponds to a vapour pressure of 1.29 kPa at point C.

The calculation of the vapour pressure distribution proceeds in the same manner as for the temperature distribution (see *Table 3*). Note that the vapour resistances at the surfaces are small and can be ignored.

It is now necessary to determine the dew point temperatures corresponding to the vapour pressures. The vapour pressures concerned are 1.29 kPa for the inside air, 1.286 kPa for the insulation-brick interface and 0.65 kPa for the brick-outside air interface. The dew point temperature is found by drawing horizontal lines through these values until they cut the 100% relative humidity curve and then reading downwards to the temperature scale. These lines are shown on *Fig 3*, the lines for 1.29 kPa and 1.286 kPa are so close that they are shown as one line. The line for 0.65 kPa is the line AB produced backwards. It will be seen that the corresponding dew point temperatures are 10.7°C and 0.8°C. The dew point temperature distribution is shown on *Fig 2*.

Condensation will occur when the value of the temperature distribution is below the value of the dew point temperature distribution. The shaded region in *Fig 2* shows where this occurs, and condensation is thus likely to occur in the region of the interface between the insulation and the brickwork.

To reduce the likelihood of interstitial condensation one possible method is to include a vapour barrier in the construction. The next problem will illustrate that the position of the vapour barrier must be carefully selected. The construction to be considered is that given in *Problem 2*, and is used to demonstrate the principles even though the construction may not be entirely practical.

Problem 3 A vapour barrier consisting of sheet polythene is to be incorporated in the construction detailed in *Problem 2*. The polythene has a thickness of 0.4 mm, a thermal conductivity of 0.35 W/mK and a vapour resistance of 200×10^9 Ns/kg.

Determine the likelihood of interstitial condensation if the vapour barrier is placed
(i) between the insulating board and the brickwork;
(ii) on the surface of the insulating board. Draw relevant conclusions.

It would appear necessary to recalculate the temperature distribution in each case but the thermal resistance of the polythene is very small. The actual thermal resistance is $0.0004/0.35 = 0.001$ m^{2}K/W which is so insignificant compared with the other thermal resistances found in *Problem 2* that its effect on the temperature distribution can be ignored. The temperature distribution found in *Problem 2* will be used. The vapour pressure distribution will need to be calculated in each case, see *Tables 4 and 5*.

TABLE 4

Element	Vapour resistance (Ns/kg)	Vapour pressure drop (kPa)	Vapour pressure (kPa)	Dew point temp. (°C)
Insulation	0.1×10^9	$\dfrac{0.1 \times 10^9}{215.1 \times 10^9} \times 0.64 \approx 0.0$	1.29	10.7
Vapour barrier	200×10^9	$\dfrac{200 \times 10^9}{215.1 \times 10^9} \times 0.64 = 0.595$	0.695	1.5
Brick	15×10^9	$\dfrac{15 \times 10^9}{215.1 \times 10^9} \times 0.64 = 0.045$	0.65	0.8
Totals	215.1×10^9	0.64		

Notes. (i) The vapour resistances for the insulation and the brick are taken from *Problem 2*. The vapour resistance of the polythene is given.
(ii) The vapour pressures and dew point temperatures were found using the same method as in *Problem 2*.

TABLE 5

Element	Vapour resistance (Ns/kg)	Vapour pressure drop (kPa)	Vapour pressure (kPa)	Dew point temp (°C)
Vapour barrier	200×10^9	0.595	0.695	1.5
Insulation	0.1×10^9	0.0	0.695	1.5
Brick	15×10^9	0.045	0.65	0.8
Totals	215.1×10^9	0.64		

The vapour pressure drops are calculated as in case (i) and the vapour pressures are found by successive subtraction from the inside vapour pressure of 1.29 kPa

Case (i)
The appropriate temperature distributions are shown in *Fig 4*. It will be seen that condensation is still likely to occur in the insulating material.

Case (ii)
The temperature distributions are shown in *Fig 5*. It will be seen that there is now no likelihood of condensation occurring within the structure. This illustrates the fact

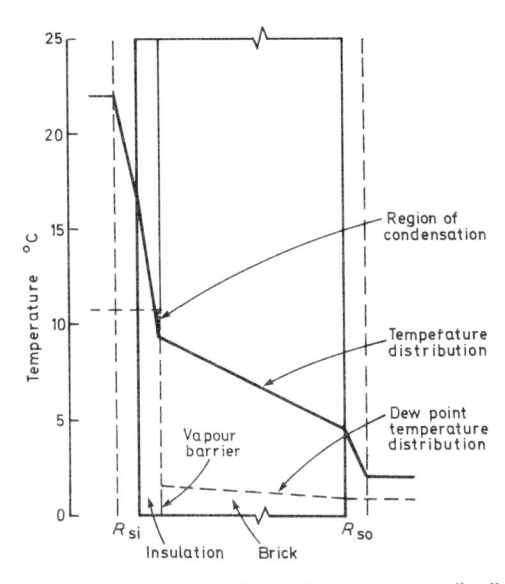

Fig 4 Temperature and dew point temperature distributions for Problem 3

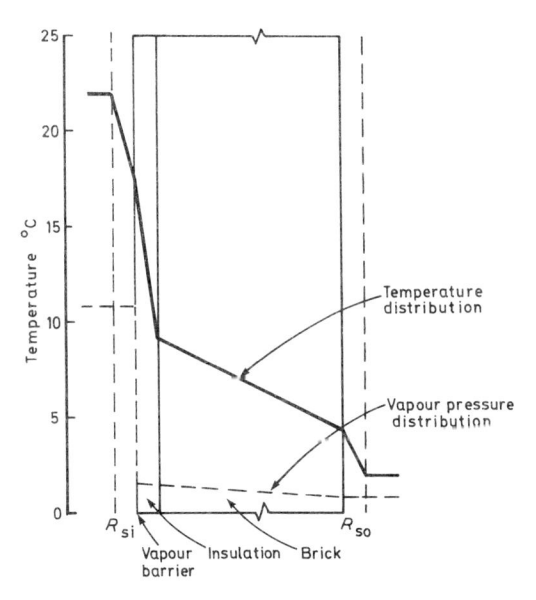

Fig 5 Temperature and dew point temperature distributions for Problem 4

that the vapour barrier should be placed as near the inside of the construction as is practically possible.

It is possible to reduce the likelihood of condensation by reducing the moisture excess inside the building. One possible method is to reduce the amount of water vapour released in the building. Increased ventilation may reduce the amount of water vapour present but will naturally reduce the internal air temperature for a given heat input. Specific measures to remove water vapour from cooking, clothes drying, bathrooms and other wet processes may be beneficial.

EXERCISES (answers on page 184)

1. A single glazed window has a U-value of 5.7 W/m² K. The inside and outside air temperatures are 18°C and 1°C respectively. Assuming that the inside surface resistance is 0.12 m² K/W determine:
   (i) the inside surface temperature of the glass;
   (ii) the relative humidity of the inside air at which condensation will begin to form on the inside of the window.

2. A window has a U-value of 5.7 W/m² K. The outside air has a temperature of 2°C and a 90% relative humidity. The inside air has a temperature of 18°C and a moisture excess of 0.004 kg/kg dry air. If the inside surface resistance is 0.12 m² K/W determine:
   (i) the inside surface temperature of the glass
   (ii) the vapour pressure of the inside air;
   (iii) whether condensation will occur on the inside of the glass.

3. Determine whether condensation will occur in the inner leaf of the cavity wall construction shown in *Fig 6*, which also shows the inside and outside air conditions. The following data should be used:
   Thermal resistances: inside = 0.12 m² K/W
   cavity = 0.18 m² K/W
   outside = 0.06 m² K/W

INSIDE CONDITIONS
temperature = 20°C
vapour excess = 0.004 kg/kg dry air

OUTSIDE CONDITIONS
Temperature = 0°C
relative humidity = 100%

Plasterboard

Insulation   Brick   Cavity   Brick

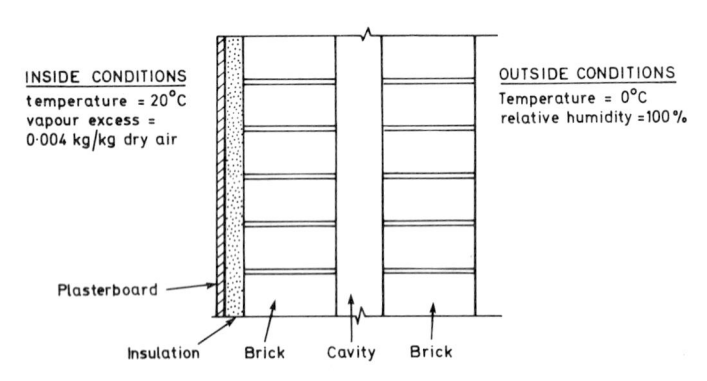

Fig 6 Construction for Exercise 3

Material	Thickness (m)	Thermal conductivity (W/mK)	Vapour resistivity (Ns/Kg m)
Plasterboard	0.010	0.16	60×10^9
Insulation	0.020	0.035	0
Brick	0.105	0.84	50×10^9

4 Repeat *Exercise 3*, using the same data, for the modified construction shown in *Fig 7*.

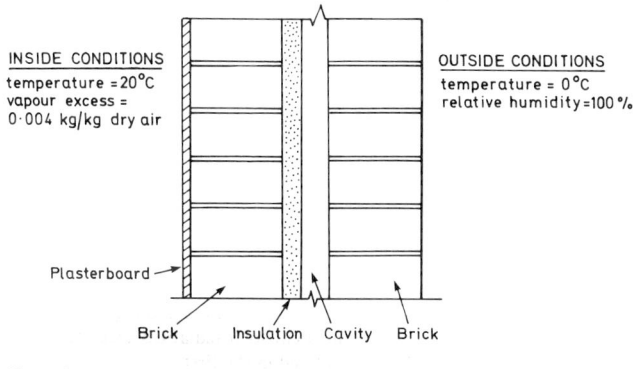

Fig 7 Construction for Exercise 4

5 Select the correct option: in a construction the vapour barrier should be placed:
   (a) as near the outside as possible;
   (b) between any insulating material and the outside;
   (c) as near the inside as possible;
   (d) wherever convenient.

4 Solar and casual heat gains

SOLAR GAINS

Heat gain due to solar radiation is most important with lightweight buildings having large areas of glazing. In the design of a building it may be necessary to calculate either the maximum cooling load on the air conditioning system or the peak temperatures encountered in a non air conditioned building.

The detailed calculations are somewhat complex and are beyond the scope of this book but an understanding of some of the factors is vital both to the above-mentioned problems and to the utilisation of solar radiation for heating systems. As the maximum cooling load or the peak temperature will occur after several days of clear sunshine these conditions are used in the design procedure.

One major factor is the amount of solar radiation striking the glazing that is transmitted to the inside of the building, thus the incident solar radiation and the transmission characteristics of the glass will be considered in the first instance.

PROPERTIES OF THERMAL AND SOLAR RADIATION

All bodies emit radiation, the intensity of the radiation and its wavelength distribution depend upon its temperature as illustrated in *Fig 1*. The curves shown are for black body radiators. The temperatures are given in degrees absolute, which are obtained by adding 273 to the temperature in degrees Celsius; for example $20°C = 293°K$. It will be seen that the higher the temperature the greater the intensity of the radiation and the shorter the wavelength at which the peak intensity occurs. The wavelength, λ_{max}, at which the maximum occurs is related to the temperature, T, by Wien's displacement law:

$\lambda_{max} T = 2897.8 \ \mu m°K$

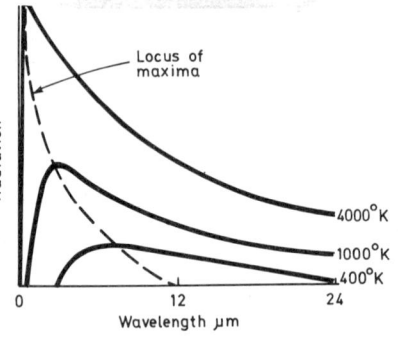

Fig 1 Wavelength distribution for a black body radiator

Problem 1 Find the wavelength of the maximum intensity for a body at (i) $6000°K$; (ii) $30°C$.

(i) Using the above formula with $T = 6000$:
$\lambda_{max} \times 6000 = 2897.8$

$\lambda_{max} = \mathbf{0.483\ \mu m}$

(ii) In this case $T = 30 + 273 = 303°K$
Hence: $\lambda_{max} \times 303 = 2897.8$

$\lambda_{max} = \mathbf{9.56\ \mu m}$

It will be noted that for high temperatures the wavelength at the maximum is small and such radiation is referred to as short wavelength radiation and is similar to the radiation from the sun outside the earths atmosphere.

At low temperatures long wave radiation is produced which is typical of the radiation from a warm building surface. The importance of this will become apparent when the transmission characteristics of glass are considered.

The Earth's atmosphere absorbs some of the solar radiation and at ground level it is found that most of the solar energy is in the wavelength range 0.29 to 3.0 μm. This includes the visible spectrum of 0.38 to 0.78 μm.

COMPONENTS OF SOLAR RADIATION

The solar radiation reaching a surface has three components:
(i) direct radiation,
(ii) diffuse sky radiation, and
(iii) radiation reflected from the ground.

The intensity of the direct radiation depends on the latitude, the time of year, the time of day, the orientation of the surface and the inclination of the surface.

Solar radiation passing through the atmosphere is scattered and thus creates a source of diffuse radiation. The amount of diffuse radiation depends on the same factors as direct radiation except that it is taken to be independent of the orientation of the surface. Solar radiation reflected from the ground creates diffuse radiation which will impinge on vertical surfaces. It is usual to assume that the reflection factor of the ground is 0.2 except for arid tropical zones.

The diffuse radiation may form a significant part of the total radiation incident upon a building surface. *Fig 2* shows a comparison between the direct radiation and the diffuse radiation incident upon a vertical, South-east facing wall in April; the latitude being 50°N.

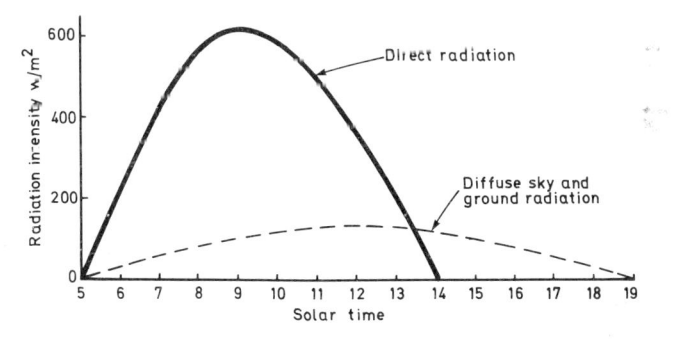

Fig 2 Direct and diffuse solar radiation on a south-east facing wall in April

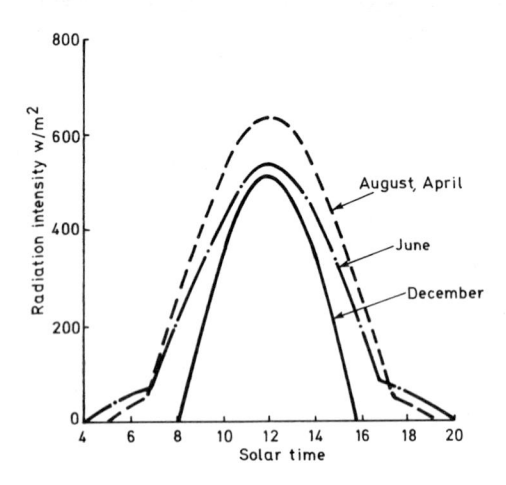

Fig 3 Total solar radiation on a South facing wall

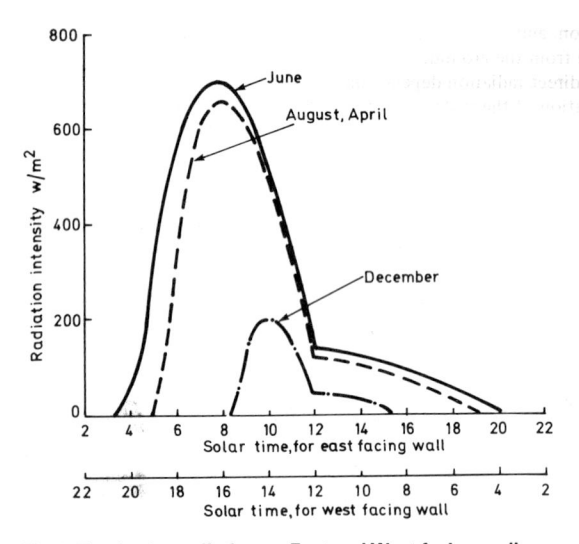

Fig 4 Total solar radiation on East and West facing walls

For many purposes only the total amount of radiation falling on the building surface needs to be known. *Figs 3, 4 and 5* illustrate the total radiation on a vertical wall for selected orientations and times of the year. All these diagrams are applicable for a latitude of 51.7°N.

It will be seen from *Fig 3* that the values are symmetric about solar noon. In *Fig 4* the radiation values for West facing walls can be obtained from those for East facing

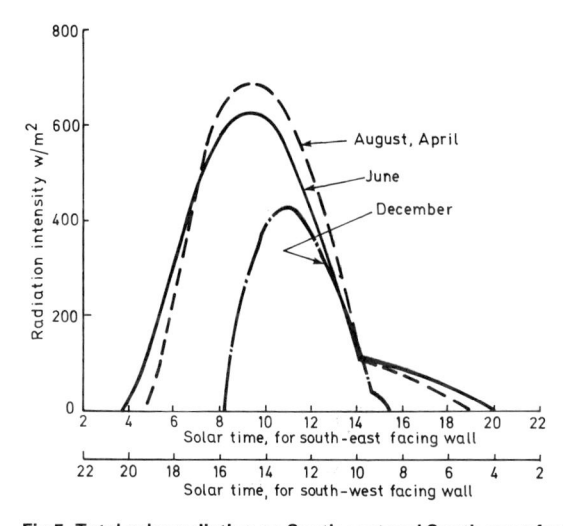

Fig 5 Total solar radiation on South-east and South-west facing walls

walls by reversing the time scale, the same situation holds for South-east and South-west walls in *Fig 5*.

TRANSMISSION CHARACTERISTICS OF GLASS

The amount of radiation transmitted through glass depends on the wavelength of the radiation. *Fig 6* shows a slightly simplified transmission curve for glass. It will be seen that the transmittance is high for wavelengths from 0.3 to 3.0 μm which is the wavelength range for solar radiation.

The transmittance for wavelengths over 3μm is very small and glass is thus opaque to long wavelength radiation which is characteristic of low temperature sources. Thus solar radiation will pass through glass and will strike internal surfaces which will be warmed up. These surfaces will then re-radiate long wavelength radiation which will not pass out through the glass. This is the explanation of the 'greenhouse' effect.

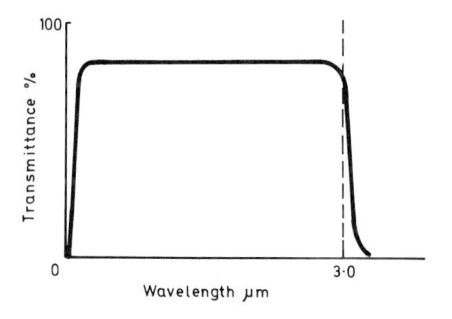

Fig 6 Simplified transmission characteristics of clear glass

The direct solar radiation which strikes the glass is partially reflected, partially absorbed and the remainder is directly transmitted. Thus:

direct transmittance + reflectance + absorptance = 1

The values of the direct transmittance, reflectance and absorptance depend upon the angle of incidence of the direct radiation and typical values for clear glass are shown in *Fig 7*. It will be noted that for large angles of incidence the reflectance increases rapidly whereas the direct transmittance and absorptance decrease.

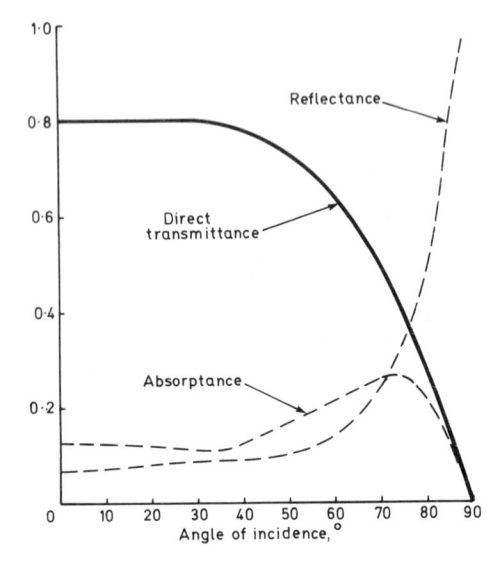

Fig 7 Variation of solar reflectance, transmittance and absorptance with angle of incidence

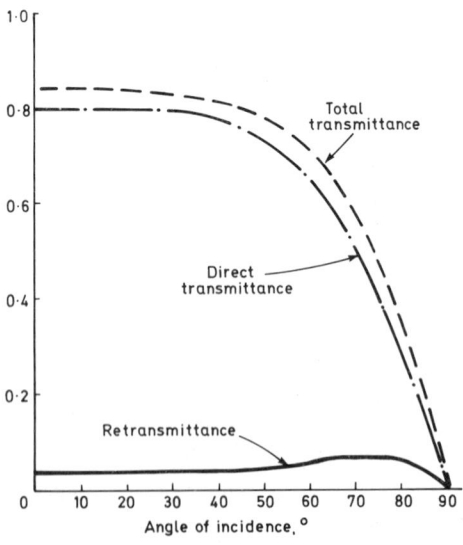

Fig 8 Direct transmittance and retransmittance for clear single glazing

The direct transmittance curve in *Fig 7* does not represent the complete picture. The solar radiation absorbed by the glass causes an increase in its temperature. The glass then retransmits the absorbed radiation partially to the inside of the building and the remainder to the outside. For single glazing 30% of the absorbed radiation is retransmitted to the inside. The total transmittance is the sum of the direct transmittance and retransmittance. *Fig 8* illustrates the total and direct transmittances for clear glass.

> *Problem 2* At a particular angle of incidence the absorptance and direct transmittance for solar radiation of a glass used for solar control are 0.36 and 0.54 respectively. Determine the reflectance and total transmittance for solar radiation when the glass is used as single glazing.

By rearrangement of the equation given above:
reflectance = 1 − (absorptance + direct transmittance)
= 1 − (0.36 + 0.54) = **0.10**
From above the retransmittance is 30% or 0.3 of the absorptance. Hence:
total transmittance = direct transmittance + retransmittance
= 0.54 + (0.3 × 0.36) = **0.65**
It is instructive to represent the transmitted, reflected and absorbed radiation diagrammatically, this has been done in *Fig 9*, taking the incident radiation as 100.

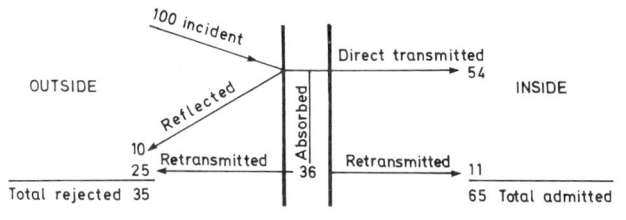

```
                100 incident
                     \         |  Direct transmitted
                      \        |         → 54
OUTSIDE                \       |                         INSIDE
             Reflected  \    Absorbed
              10 ←       \     |
              25   Retransmitted    Retransmitted
                              36         → 11
Total rejected 35                        65 Total admitted
```

Fig 9 Transmission characteristics of glass in Problem 2

Diffuse radiation is incident upon the glass from all angles and average values of the absorptance and transmittance may be used. For clear glass approximate values of the diffuse absorptance and diffuse transmittance would be 0.2 and 0.67 respectively. The retransmittance for diffuse radiation for single glazing can be calculated in the manner given in *Problem 2*.

> *Problem 3* The solar radiation incident upon a vertical window at a particular time had a direct intensity of 600 W/m² and a diffuse sky and ground intensity of 72 W/m². The solar properties of the single glazing are:
> direct transmittance 0.72   diffuse transmittance   0.67
> direct absorptance   0.19   diffuse absorptance   0.2
> Evaluate: (i) direct total transmittance; (ii) diffuse total transmittance;
>   (iii) total solar heat gain.

(i)   direct total transmittance = 0.72 + (0.3 × 0.19) = **0.78**
(ii) diffuse total transmittance = 0.67 + (0.3 × 0.2) = **0.73**
(iii) the total solar heat gain is found by multiplying the direct and diffuse intensities by the appropriate total transmittances found above and adding the results.
total solar heat gain = 600 × 0.78 + 72 × 0.73
= **521 W/m²**

METHODS OF REDUCING SOLAR GAIN THROUGH WINDOWS

Solar controls achieve a threefold purpose in that they reduce visual glare, prevent direct sunshine falling on the occupants, and diminish the solar radiation gain to the interior of the building. The following are the main types of solar controls:

(a) *Internal.* These give good protection against visual glare and direct sunshine. Internal blinds absorb radiation causing the blind to become warm. The blind then retransmits heat mainly to the inside of the room. Thus the effectiveness of internal blinds in reducing peak temperatures and cooling loads is reduced. For best efficiency the blinds should be light coloured to assist in reflecting radiation.

(b) *Solar control glasses.* These can be approximately classified as solar absorbing glasses and solar reflective glasses. The solar absorbing glasses are produced in a wide range of colours including green, grey and bronze. In all cases the solar reflectance is small, 0.05 being typical. The solar absorptance varies widely depending upon density of the colour and varies from about 0.35 to 0.75. The total transmittance varies from about 0.45 to 0.70. The light transmittance of these glasses is less than that of clear glass and depends on the colour and colour density.

The solar reflective glasses are also produced in a range of colours including gold, silver, bronze and silver blue. The solar reflectance is higher than for the solar absorbing glasses and ranges from about 0.25 to 0.60. The total transmittance again varies widely depending on the particular glass but is typically in the range 0.15 to 0.55. The light transmittance of these glasses is variable but typically in the range 0.15 to 0.40.

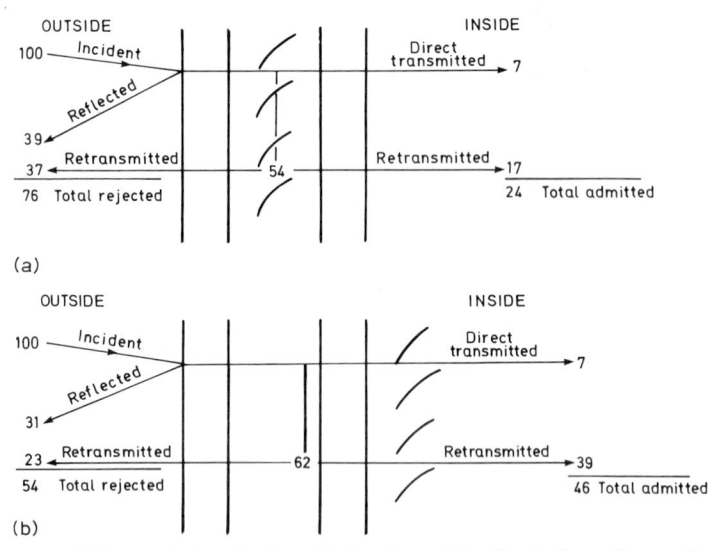

Fig 10 (a) Transmission of solar radiation through double glazing with venetian blind between panes.
(b) Transmission of solar radiation through double glazing with venetian blind inside

(c) *External.* These are the most effective since the absorbed radiation is retransmitted externally to the building. This type of solar control includes shutters, awnings and projecting horizontal or vertical fins which may be part of the construction. Because of the low sun angles in the UK horizontal projections are less efficient than vertical systems.

(d) *Sun controls between panes of double glazing.* The use of a blind between panes of double glazing is more efficient than the use of the same blind inside the double glazing. This arises since the retransmitted heat is dissipated less to the inside of the building when the blind is between the panes.

Figs 10(a) and (b) illustrate this comparison. In *Fig 10(a)* it will be seen that 54% of the incident radiation is absorbed by the double glazing and the blind. Only 17% is retransmitted inwards to give a total transmittance of 24%. In *Fig 10(b)*, 62% of the incident radiation is absorbed by the double glazing and the blind. In this case since the blind is on the inside 39% is retransmitted inwards to give a total transmittance of 46%. Thus the blind between the panes of glazing is considerably more efficient and will have the advantage of keeping the blinds clean and thus reducing maintenance costs.

In double glazing systems it may be advantageous to use a solar control glass for the outer panes.

Casual gains

In addition to solar gain a number of other sources of casual heat gain exist in buildings. For precise estimation of the heat gains considerable information is required but the following will illustrate some of the sources to be considered.

Human bodies

As already stated in chapter 1 the human body looses heat continuously. About 75% is lost by convection and radiation and about 25% is lost by evaporation. In allowing for heat gains to the building only that heat loss from the human body by convection and radiation need be considered; this is termed the sensible heat loss.

TABLE 1 Sensible heat loss (W) at different air temperatures

Activity	Sensible heat loss at given air temperature (°C)		
	15°C	20°C	25°C
Seated at rest	100	90	70
Light work	110	100	75
Medium work	160	140	97
Heavy work	220	190	120

The sensible heat loss depends upon the environment and the level of activity of the person. *Table 1* gives some figures for human males.

> *Problem 4* Estimate the annual casual gain during the heating season from two adults and a child of school age in a dwelling.

There is no precise answer to this question since it depends on the assumptions made and the following assumptions are reasonable:
(i) Sensible heat emission from adult male in domestic situation is 100 W. The heat emission from a woman and a child is less than this probably 85 W and 75 W respectively.
(ii) Each person is in the dwelling for 16 hours a day on average.
(iii) The heating season is 180 days.
The total heat gain rate = 260 W
Heat gain in one hour = 260 × 3600 J = 936 000 J
Heat gain in one day   = 936 000 × 16 J = 14.98 MJ
Heat gain in the season = 14.98 × 180 MJ = **2.7 GJ**.

Lighting

Virtually all the energy consumed by the luminaires will be converted to heat within the building. For an illuminance of 500 lux, which is suitable for routine work in many situations, the energy load could be as high as 30 watts per square metre of floor area. This figure assumes that appropriate equipment is used and would be considerably higher if, for instance, a large office were lit with tungsten filament lamps.

The distribution of this heat within the building will depend on the type of luminaire. For pendant luminaires all the energy supplied will enter the occupied space. For luminaires fixed directly to the ceiling about 80% of the heat enters the occupied space and the remainder enters the ceiling. For recessed liminaires the amount of heat entering the occupied space reduces to about 50%, the remainder entering the ceiling void.

Heat from the lamps is lost by convection, conduction and radiation. The amount of radiation is greatest for tungsten filament lamps and least for fluorescent lamps. The type of luminaire in which the lamp is installed has a significant effect on the proportions of heat lost by convection, conduction and radiation.

Other sources

Heat gains will occur from many other sources these include electrical appliances such as cookers, televisions, motors, data processing equipment and gas appliances such as cookers and water heaters. The CIBS guide gives figures of typical heat outputs.

EXERCISES (answers on page 185)

In the following questions where options are given select the correct option or options.

1 Solar radiation:
   (a) is short wavelength radiation; (b) is long wavelength radiation.
   (c) has the majority of its energy at wavelengths less than 3 μm.
   (d) has the majority of its energy at wavelengths greater than 3 μm

2 The wall of a building is at a temperature of 25°C. The energy radiated by the wall has a maximum intensity at a wavelength of:
   (a) 115.9 μm; (b) 9.7 μm; (c) 3 μm.

3 On clear days the maximum intensity of solar radiation on a vertical wall facing South is:
   (a) greatest in December; (b) greater in June than April; (c) greater in April than June.

4 On clear days the maximum intensity of solar radiation on a West facing vertical wall occurs:
   (a) at mid-day; (b) during the morning; (c) during the afternoon.

5 On clear days the maximum intensity of solar radiation on an East facing vertical wall occurs:
   (a) at mid-day; (b) during the morning; (c) during the afternoon.

6 Clear glass is:
   (a) opaque to solar radiation;   (c) opaque to long wave radiation;
   (b) transparent to solar radiation;   (d) transparent to long wave radiation.

7 At a particular angle of incidence the absorptance and direct transmittance for solar radiation of a glass are 0.29 and 0.43 respectively. Determine the reflectance and total transmittance when the glass is used as single glazing.

8 A double glazing system is to have one pane of solar absorbing glass and one of clear float glass. The solar absorbing glass should be used for:
   (a) the inner pane; (b) the outer pane;
   because:
   (c) the solar absorbing glass absorbs more heat than the clear glass.
   (d) the clear glass absorbs more heat than the solar absorbing glass and can thus retransmit heat mainly to the
   (e) outside; (f) inside.

9 The solar radiation incident upon a vertical window at a particular time had a direct intensity of 480 W/m^2 and a diffuse sky and ground intensity of 115 W/m^2. The solar properties of the single glazing are:

   direct transmittance   0.66   diffuse transmittance   0.67
   direct absorptance   0.21   diffuse absorptance   0.2

   Calculate (i) the direct total transmittance; (ii) the diffuse total transmittance; (iii) the total solar heat gain.

10 An internal blind:
   (a) gives good protection from direct sunshine falling on occupants;
   (b) retransmits absorbed radiation mainly to the inside;
   (c) retransmits absorbed radiation mainly to the outside;
   (d) should be light coloured to absorb as much radiation as possible;
   (e) should be light coloured to reflect as much radiation as possible.

11 Estimate the casual heat gain rate due to the occupants in a restaurant seating 80 people. It may be assumed that 40% are men, 50% are women and the remainder children. The sensible heat loss for men may be taken as 110 W and that the sensible heat loss for women and children is 85% and 75% of this figure respectively. Allow 10 W per person for heat gain from hot food.

5 Sound propagation and units of measurement

SOUND WAVES

Sound waves are a particular form of elastic waves. Elastic waves can occur in a medium which possesses both mass and elasticity. If the medium has mass then a displaced particle can transfer momentum, and hence energy to an adjacent particle. The elasticity of the medium tends to return the displaced particle to its original position.

In many cases the medium through which sound is propagated is air. Air certainly possess mass since its density is approximately 1.25 kg/m^2. The elasticity of air is simply demonstrated by a bicycle tyre pump. If a finger is placed over the outlet of the pump and the handle pushed in and then released, the handle returns towards its original position.

The propagation of a sound wave can be understood by considering a piston moving backwards and forwards in a tube of air. In *Fig 1(a)* the piston is at rest and the spacing of the air molecules is uniform along the tube. If the piston moves forward, as in *Fig 1(b)*, the air molecules adjacent to the piston are compressed, and energy is stored

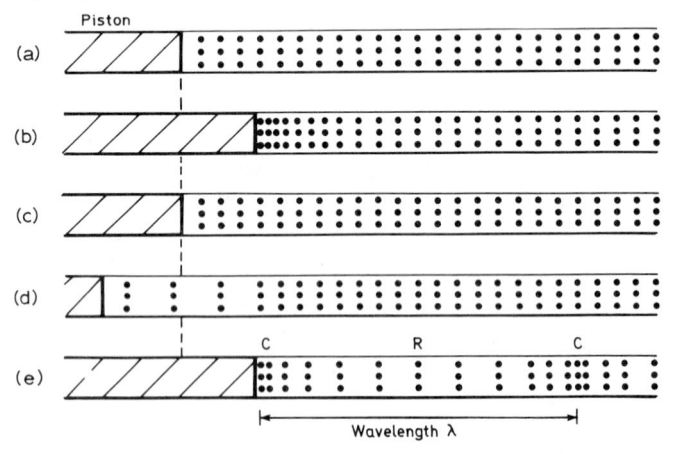

Fig 1 Propagation of a sound wave

in the compressed air. Due to the elasticity of the air the molecules endeavour to regain their original spacing. They cannot move back, since the piston is still there, so they move to the right along the tube. In doing this they acquire momentum and strike further molecules, imparting momentum to these molecules and causing them to become compressed. In this way the compression travels to the right along the tube. Suppose that the piston now returns to its original position then the air molecules near the piston assume their original spacing as shown in *Fig 1(c)*.

If the piston now moves to the left of its original position the separation of the molecules is increased and a rarefaction is created, as shown in *Fig 1(d)*. The molecules endeavour to regain their original spacing and a rarefaction is propagated to the right.

If the piston moves backwards and forwards about its original position then an alternate series of compressions and rarefactions will be propagated down the tube as shown in *Fig 1(e)* in which the letters C and R denote compressions and rarefactions respectively.

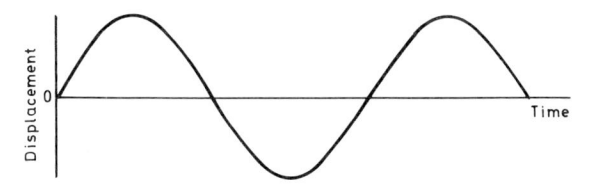

Fig 2 Displacement of an air molecule with time

The air molecules do not travel down the tube but oscillate about their original position. *Fig 2* shows the displacement of an air molecule with time when the piston oscillates.

VELOCITY OF SOUND IN AIR

The velocity of sound in air is given by

$$v = \sqrt{\frac{1.4P}{\rho}}$$

where v = velocity of sound in air in m/s;
P = the air pressure in Pa;
ρ = density of air in kg/m^3.

Problem 1 Calculate the velocity of sound in air when the atmospheric pressure is 101.3 kPa and the density of the air is 1.293 kg/m^3.

$$v = \sqrt{\frac{1.4 \times 101.3 \times 10^3}{1.293}} = \mathbf{331.2 \text{ m/s}}$$

Since temperature effects the density of air, the velocity of sound varies with temperature. The following approximate relationship holds for normal temperatures:
$v = 331.3 + 0.6t$, where t is the air temperature in °C.

Problem 2 Find the velocity of sound in air at 20°C.

From the above equation: $v = 331.3 + 0.6 \times 20 = \mathbf{343.3 \text{ m/s}}$

Problem 3 A person at an outdoor event is 10 m from one loudspeaker of the public address system and 25 m from another loudspeaker. If both loudspeakers produce the same sound at the same instant of time, calculate:
(i) the time taken for the sound to reach the person from the nearest loudspeaker.
(ii) the time delay before the sound reaches the person from the second loudspeaker.
Assume the velocity of sound in air to be 340 m/s.

(i) Remembering that:

$$\text{time} = \frac{\text{distance}}{\text{velocity}} = \frac{10}{340} = 0.0294 \text{ s}$$

(ii) The sound from the second loudspeaker has to travel an additional 15 m, so that the time delay is:

$$\text{time} = \frac{15}{340} = 0.0441 \text{ s}$$

VELOCITY OF SOUND IN LIQUIDS AND SOLIDS

The velocity of sound in liquids and solids is much greater than the velocity in air. Some typical values are:

water at 20°C	1484 m/s;
concrete	4250 to 5250 m/s;
steel	5900 to 6100 m/s.

VELOCITY, FREQUENCY AND WAVELENGTH

Referring again to *Fig 1*, the number of times in one second the piston completes a full cycle of movement is called the frequency which is denoted by f and has the unit Hertz (Hz). An example of a complete cycle of movement is from the extreme hand position as shown in *Fig 1(b)* through the positions shown in *Figs 1(c) and 1(d)* and back to the position in *Fig 1(b)*.

The wavelength of the sound is the distance between two successive compressions, or rarefactions as shown in *Fig 1(e)* and denoted by λ. The relationship between the velocity, frequency and wavelength is:

$v = f\lambda$, where v = velocity, f = frequency and λ = wavelength.

Problem 4 Find the wavelength of a sound having a frequency of 1000 Hz if the velocity of sound in air is 340 m/s.

Re-arranging the above formula: $\lambda = \frac{v}{f} = \frac{340}{1000} = 0.34 \text{ m}$

SOUND POWER, SOUND INTENSITY AND SOUND PRESSURE

The sound power of a source can be specified by the energy radiated as sound in watts. The loudest voice would produce about 0.001 w, noisy construction plant might produce 100 W and a jet aircraft might produce 100 000 W of sound energy when taking off.

The sound intensity at a distance from a source is measured in watts per square metre (W/m²). The intensity will depend on the power of the source, the distance from the source and the pattern of radiation. Several simple cases can readily be understood:

(i) if a point source radiates uniformly then at a distance r from the source the energy will be spread over the surface of a sphere of area $4\pi r^2$. Thus the intensity, I, is given by

$$I = \frac{W}{4\pi r^2}, \text{ where } W = \text{sound power in watts}$$

(ii) if a point source is at ground level, and assuming that the ground is perfectly reflecting, then at a distance r from the source the energy will be spread over the surface of a hemisphere of area $2\pi r^2$. In this case:

$$I = \frac{W}{2\pi r^2}$$

(iii) for a line source, such as a busy road, the energy will be spread over the surface of a cylinder having an area of $2\pi r$ for each metre run of the source.

It should be noted that the above considerations only apply if the source is radiating into free space. If the source is enclosed in a room the walls of which reflect sound energy the intensity will then depend on the properties of the room as well as the properties of the source.

> *Problem 5* A point source situated at ground level has a sound power of 10 W. Calculate the sound intensity at a distance of 5 m.

In this case the energy is spread over a hemisphere so that:

$$I = \frac{W}{2\pi r^2} = \frac{10}{2\pi(5)^2} = 0.064 \text{ W/m}^2$$

Sound intensity is difficult to measure; most acoustic measuring devices respond to the variation in pressure occurring in the sound wave. For plane progressive waves and spherical waves at some distance from the source it can be shown that:

intensity α (pressure)²

The intensity and pressure are related by the formula:

$$I = \frac{p^2}{\rho v}$$

where p is the sound pressure, ρ is the density and v is the velocity of sound. For air, under normal conditions, $\rho v = 410$. Since the pressure in a sound wave is varying with time, a sound wave being an alternate set of compressions and rarefactions creating positive and negative pressures, the value of the pressure used is the root mean square (r.m.s.) pressure. This is identical in concept to the r.m.s. values used in alternating current electricity.

> *Problem 6* Determine the r.m.s. pressure in a sound wave which has an intensity of 10^{-12} W/m².

On rearrangement of the above formula:
$p = \sqrt{I\rho v}$ where $\rho v = 410$
Thus $p = \sqrt{410 \times 10^{-12}} = 2.02 \times 10^{-5}$ **Pa**

DECIBEL SCALES

In selecting a scale for measuring sound it is useful to select a method which reflects the way in which the human ear responds to sound.

Two aspects need to be considered. Firstly there is a minimum intensity, termed the threshold of hearing, which the ear can detect and this has an approximate value of 10^{-12} W/m² at 1000 Hz. The intensity of the threshold of hearing varies with frequency as shown in *Fig 3*. The range of sound intensity that the human ear can appreciate is truly astonishing; the most powerful sound which is likely to be encountered has an intensity which is a factor of ten million million above the threshold.

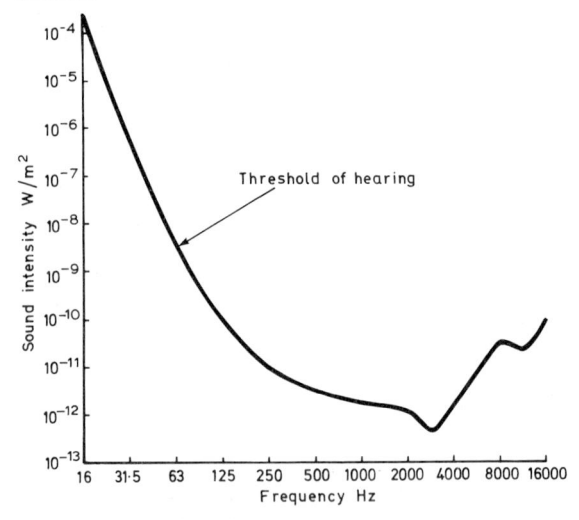

Fig 3 Threshold of hearing

Secondly the Weber-Fechner law suggests that the response of the ear is logarithmic in character which implies that the sensation produced is proportional to the logarithm of the intensity.

Thus the scale used to put objective numbers to sound intensities is logarithmic and is called the **decibel** scale. All scales must have a starting point; with an ordinary ruler, for measuring length, this is zero. Unfortunately the logarithm of zero is undefined so zero intensity cannot be used for the start of the scale. The value of 10^{-12} W/m², which is the threshold of hearing, makes a suitable starting point.

This idea is not difficult as can be seen by the following analogy. Suppose one has a ruler with a broken end so that the first mark is 20 mm. If this ruler is used to measure a line as shown in *Fig 4* then the reader will readily see that the length of the line is 80−20 = 60 mm. In the decibel scale for intensity the logarithm of the intensity is used, thus

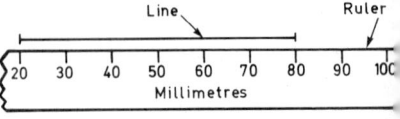

Fig 4 Ruler without a zero

intensity level, $L_I = 10\,(\log_{10} I - \log_{10} I_o)$ dB where I_o is the starting value of 10^{-12} W/m² and the factor of 10 produces a scale of suitable range, up to about 140, and generally avoids the use of decimals of a decibel.

By the use of the laws of logarithms the above formula is usually written as:

$$L_I = 10 \log_{10} \frac{I}{I_o} \text{ dB}$$

Since the intensity is related to the pressure by the formula $I = p^2/\rho v$ a similar scale for the sound pressure level, L_p can be defined:

$$L_p = 10 \log_{10} \frac{p^2}{p_o^2} \text{ dB}$$

where p_o is the reference pressure of 2×10^{-5} Pa. This is chosen so that the intensity level and sound pressure level have the same numerical value for most practical purposes. In some instances 2×10^{-5} Pa is expressed as 20 μPa.

It is useful in some circumstances, as will be seen later, to express the sound power of a source in decibels. The corresponding definition of sound power level, L_w, is

$$L_w = 10 \log_{10} \frac{W}{W_o}$$

where W_o is the reference power of 10^{-12} watts.

Care has to be taken with the reference values since in other uses of the decibel scale different reference values are used.

Problem 7 Find the sound pressure level for a sound having an r.m.s. pressure of 4×10^{-2} Pa.

From the definition of sound pressure level:

$L_p = 10 \log_{10} \dfrac{p^2}{p_o^2}$ where $p_o = 2 \times 10^{-5}$ Pa

$L_p = 10 \log_{10} \dfrac{(4 \times 10^{-2})^2}{(2 \times 10^{-5})^2}$

$= 10 \log_{10} (4 \times 10^6) = \mathbf{66 \text{ dB}}$

By the use of the laws of logarithms an alternative form of the equation for L_p can be found:

$$L_p = 10 \log_{10} \frac{p^2}{p_o^2} = 20 \log_{10} \frac{p}{p_o}$$

This form can be useful but must be treated with caution when sound pressure levels are added or subtracted.

Problem 8 A sound source produces 70 W of acoustic power; express this as a sound power level with a reference power of 10^{-12} W.

$$L_w = 10 \log_{10} \frac{W}{W_o} = 10 \log_{10} \frac{70}{10^{-12}} = \mathbf{138 \text{ dB}}$$

ADDITION AND SUBTRACTION OF DECIBELS

Since the decibel scale is logarithmic the decibel values cannot be added or subtracted directly. Only the intensities or the squares of pressures or the powers can be added or subtracted. The following problems illustrate the method.

Problem 9 Two sounds each having a sound pressure level of 60 dB are produced simultaneously. Find the resultant sound pressure level.

Applying the definition of sound pressure level to one sound:

$$L_p = 10 \log_{10} \frac{p^2}{p_o^2}$$

$$60 = 10 \log_{10} \frac{p^2}{p_o^2}$$

$$\log_{10} \frac{p^2}{p_o^2} = 6$$

By taking antilogarithms: $\frac{p^2}{p_o^2} = 10^6$

For the two sounds the square of the pressure will be doubled, so that for the resulting sound:

$$\frac{p^2}{p_o^2} = 2 \times 10^6$$

The resultant sound pressure level is then given by:

$L_p = 10 \log_{10} (2 \times 10^6) =$ **63 dB**

This illustrates the general rule that if two equal sound pressure levels are added the increase is 3 dB.

Problem 10 Determine the resulting sound pressure level obtained by adding sound pressure levels of 75 dB and 85 dB.

Applying the definition of sound pressure level to each sound in turn gives:

$$75 = 10 \log_{10} \frac{p_1^2}{p_o^2} \text{ hence } \frac{p_1^2}{p_o^2} = 3.162 \times 10^7$$

$$\text{and } 85 = 10 \log_{10} \frac{p_2^2}{p_o^2} \text{ hence } \frac{p_2^2}{p_o^2} = 3.162 \times 10^8$$

The total square pressure p^2 is given by

$$\frac{p^2}{p_o^2} = \frac{p_1^2}{p_o^2} + \frac{p_2^2}{p_o^2} = 3.162 \times 10^7 + 3.162 \times 10^8 = 3.478 \times 10^8$$

The resulting sound pressure level is found thus:

$$L_p = 10 \log_{10} \frac{p^2}{p_o^2} = 10 \log_{10} (3.478 \times 10^8) = \textbf{85.4 dB}$$

This problem illustrates the general rule that if two sound pressure levels differing by 10 dB or more are added together there is virtually no increase in level above the highest value.

Problem 11 Six similar machines running together produce a sound pressure level of 90 dB. Find the sound pressure level if four of the machines are turned off

For all six machines applying the definition of sound pressure level:

$$90 = 10 \log_{10} \frac{p^2}{p_o^2} \text{ hence } \frac{p^2}{p_o^2} = 10^9$$

For the two machines left running:
$$\frac{p^2}{p_o^2} = \frac{2}{6} \times 10^9 = 0.3333 \times 10^9 = 3.333 \times 10^8$$

The final sound pressure level is given by:
$$L_p = 10 \log_{10} \frac{p^2}{p_o^2} = 10 \log_{10} (3.333 \times 10^8) = 85.2 \text{ dB}$$

The above method can be simplified in many cases by the use of the chart shown in *Fig 5* the use of which is illustrated in the next problem.

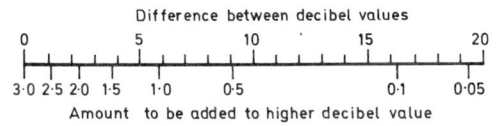

Fig 5 Scale for addition of decibels

Problem 12 Find the resultant sound pressure level obtained by adding sound pressure levels of 82 dB and 86 dB

Difference between sound pressure levels = 86 − 82 = 4 dB.
From *Fig 5*: amount to be added to higher level = 1.4 dB
  Resultant sound pressure level = 86 + 1.4 = **87.4 dB**

The foregoing problems illustrated the addition and subtraction of sound pressure levels but the methods employed can be used with any decibel scale, for instance intensity levels and sound power levels.

LOUDNESS OF PURE TONES

The loudness of a pure tone is a subjective effect which depends on the amplitude of the pressure variation in the sound wave and its frequency. In addition to the above

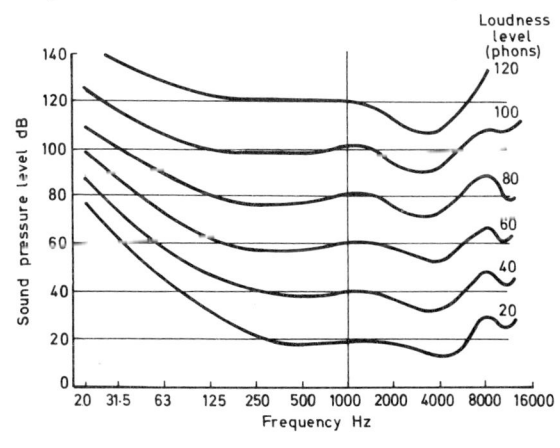

Fig 6 Equal loudness contours for pure tones

factors the sensation of loudness varies from person to person depending on their ear and brain. To obtain a scale of loudness level for people with normal hearing tones of different frequency are compared with a 1000 Hz tone to obtain equal loudness contours.

Some of these equal loudness contours are shown in *Fig 6*. The loudness level for each contour is given in phons. As an example consider the 60 phon loudness level: to obtain this curve a sound of 60 dB sound pressure level at 1000 Hz is used as a reference. For other frequencies the sound pressure level is adjusted so that the sound appears as loud as the reference sound. For example at 100 Hz a sound pressure level of 70 dB will sound as loud as 60 dB at 1000 Hz. By using a wide range of frequencies in turn the 60 phon equal loudness contour can be constructed. For each loudness level it will be seen that its value is given by the sound pressure level at 1000 Hz.

The main features of the equal loudness contours are important since they imply that the ear is less sensitive at very low and very high frequencies and most sensitive around 3000 Hz. The reader should study *Fig 6* and convince themself of the truth of this statement.

The loudness of noise is a considerably more complex phenomenon and is beyond the scope of this book.

SOUND LEVEL METERS AND WEIGHTING SCALES

A sound level meter consists basically of a microphone, an amplifier and an indicating meter. The microphone produces an electrical signal proportional to the r.m.s. sound pressure. This signal is amplified by a suitable stepped amplifier and the result displayed either on a normal moving pointer meter or a digital display.

As the response of the human ear varies with frequency attempts have been made to modify the sound pressure level reading of the sound level meter to accord more closely with subjective response. This has been done by introducing into the sound level meter electronic weighting networks. The characteristics of these networks, called A, B, C and D weightings are shown in *Fig 7*.

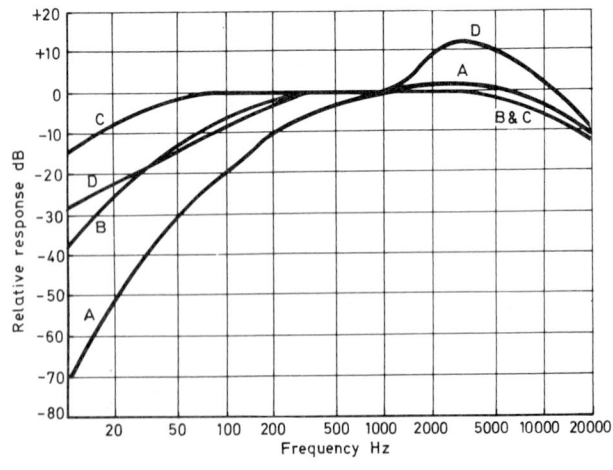

Fig 7 A, B, C and D weighting curves for sound level meters

Consider the A weighting response. At low and high frequencies the true sound pressure level value will be reduced by the amount shown on the curve in *Fig 7* to reflect the lower sensitivity of the human ear at these frequencies. This weighting is similar to the 40 phon equal loudness contour in *Fig 6*.

The B and C weightings were intended to be used as higher sound pressure levels where the equal loudness contours are flatter. The D weighting emphasises sound having a frequency around 3000 Hz where the human ear is most sensitive and is intended for assessing jet aircraft noise for which this frequency predominates.

Readings taken with these weightings are specified as dB(A), dB(B), dB(C) or dB(D). The brackets are often omitted, as for example dBA. In many circumstances the subjective response accords well with readings on the A weighting and the dBA has become widely used in noise control.

SOUND SPECTRA

Pure tones are seldom encountered; most sounds consisting of a large number of frequencies. It is usual to analyse the sound into frequency bands to give a spectrum. Octave frequency bands may be used having centre frequencies selected from the following list: 16, 31.5, 63, 125, 250, 1000, 2000, 4000, 8000 Hz. One third octave bands are also commonly used, in which the above octaves are each subdivided into three.

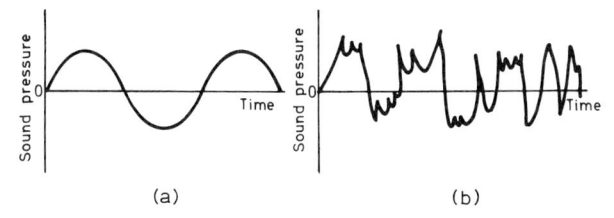

Fig 8 Waveforms of (a) a pure tone; (b) a complex sound

A possible sound spectrum for traffic noise is shown in the table below. It will be seen that the maximum levels occur at the lower frequencies. The frequency spectrum will depend on the speed of the traffic and the number of heavy vehicles.

Frequency, (Hz)	63	125	250	500	1000	2000	4000	8000
Sound pressure level (dB)	80	83	82	81	75	69	65	57

The waveform, that is the variation in sound pressure with time, of a pure tone is sinusoidal as shown in *Fig 8(a)*. For more complex sounds and noises the waveform will be more complex as shown in *Fig 8(b)*.

EXERCISES (answers on page 185)

1. Calculate the velocity of sound in air at a temperature of 15°C.

2. In a hall with a public address system a member of the audience is 17 m from the lecturer. The lecturer is using the public address system and the person is 3 m from the nearest loudspeaker. Assuming that the electrical signal from the microphone reaches the loudspaker instantaneously, find the delay between the arrival of the sound from the loudspeaker and that from the lecturer. Assume the velocity of sound in air to be 340 m/s.

3. Standing waves can be created between two parallel walls in a room if the distance between the walls is $\lambda/2, \lambda, 3\lambda/2 \ldots$, where λ is the wavelength of sound in air. For two walls separated by a distance of 3 m calculate the frequencies of the first five possible standing wave systems, assuming the velocity of sound to be 340 m/s.

4. A point source, situated at ground level has a sound power of 10 W. Calculate:
   (a) the sound intensity at a distance of 10 m;
   (b) the intensity level in decibels at a distance of 10 m using a reference intensity of 10^{-12} W.

5. A sound source has a sound power level of 100 dB, using a reference value of 10^{-12} W, find the acoustic power in watts produced by this source.

6. Three sounds each have a sound pressure level of 65 dB. Calculate the resultant sound pressure level.

7. In determining the predicted noise level due to traffic on a proposed new road it was necessary to add the following sound pressure levels: 68 dB, 72 dB and 74 dB. Calculate the total sound pressure level.

8. In a machine shop the sound presure level due to 10 identical machines is 94 dB. How many machines must be stopped in order to reduce the sound pressure level to 90 dB.

9. The background sound pressure level in a classroom is 50 dB. With 10 students working in the room the sound pressure level rises to 60 dB. Find the likely sound pressure level with 30 students in the room.

10. Two sounds each have a sound pressure level of 58 dB. Select from the options below the resultant sound pressure level.
    (a) 116 dB; (b) 63 dB; (c) 61 dB; (d) 58 dB.

11. Two sounds have sound pressure levels of 60 dB and 82 dB. Select from the options below the resulting sound pressure level, to the nearest decibel.
    (a) 142 dB; (b) 85 dB; (c) 63 dB; (d) 82 dB.

12. Six machines produce a sound pressure level of 80 dB. Select from the options below the resulting sound pressure level when three of the machines are stopped.
    (a) 77 dB; (b) 83 dB; (c) 74 dB; (d) 40 dB.

13. A sound has a sound pressure level of 60 dB and a frequency of 100 Hz. Select from the options the reading this sound would cause on a sound level meter with an A– weighting network.
    (a) 41 dB; (b) 60 dB; (c) 79 dB; (d) 54 dB

6 Sound in rooms

SOUND PRESSURE LEVEL IN A ROOM

For a sound source which radiates uniformly in all directions as shown in *Fig 1(a)* it has already been shown that:

$$I = \frac{W}{4\pi r^2}$$

where I = sound intensity, W is the power of the source and r is the distance from the source. This is often referred to as the inverse square law.

When such a source is placed in a room the sound is received by two methods: firstly direct sound and secondly reflected sound as shown in *Fig 1(b)*. The reflected

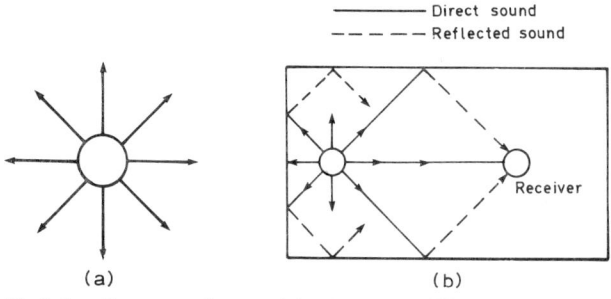

Fig 1 A uniform sound source (a) in free space; (b) in a room

sound field is usually termed the reverberant sound field. The energy in the reverberant sound field depends on the absorption properties of the surfaces of the walls. This property is measured by the absorption coefficient, denoted by α, which is defined as

$$\alpha = \frac{\text{energy absorbed}}{\text{energy incident}}$$

If the room surfaces consist of a number of materials of area $S_1, S_2, S_3 \ldots$ having absorption coefficients $\alpha_1, \alpha_2, \alpha_3 \ldots$, then an average absorption coefficient $\bar{\alpha}$, can be found as:

$$\bar{\alpha} = \frac{\alpha_1 S_1 + \alpha_2 S_2 + \alpha_3 S_3 + \ldots}{S_1 + S_2 + S_3 + \ldots}$$

$$= \frac{\Sigma \alpha_i S_i}{S} \text{ where } S = \text{total area}$$

It may be shown that the sound intensity in the reverberant field is given by:

$$I = \frac{4W}{R_c}$$

where $R_c = S\bar{\alpha}/(1-\bar{\alpha})$ and is called the room constant.

The total intensity is the sum of the intensities in the direct and reverberant fields and is given by

$$I = \frac{WQ}{4\pi r^2} + \frac{4W}{R_c} = W\left(\frac{Q}{4\pi r^2} + \frac{4}{R_c}\right)$$

The directivity factor Q has been introduced since noise sources in rooms seldom radiate energy equally in all directions. Reflecting surfaces such as floors and walls will alter the distribution of sound; the following examples will illustrate this.

If the source stands on a reflecting surface, such as a floor, the sound energy is radiated over a hemisphere and the intensity is doubled and the directivity factor, Q is 2. If the source stands at the junction of two surfaces, for example a floor and a wall, the sound now radiates over a quarter of a sphere thus the intensity is quadrupled and Q is 4. For a source in a corner of a room, that is at the intersection of three surfaces, the directivity factor is 8.

By taking logarithms of the above expression for the intensity an expression can be deduced for the sound pressure level:

$$L_p = L_w + 10 \log_{10}\left(\frac{Q}{4\pi r^2} + \frac{4}{R_c}\right)$$

Care must be exercised in the use of this equation since it will not predict the sound pressure level near to large sources and many sources, particularly machines, do not radiate sound uniformly in all directions.

> *Problem 1* A machine of sound power level 90 dB is situated against one wall of a workshop. The workshop is 30 m by 15 m on plan and 5 m high. Determine the sound pressure level at:
> (a) a distance of 5 m from the machine when the average absorption coefficient is (i) 0.1; (ii) 0.4.
> (b) a distance of 20 m from the machine when the average absorption coefficient is (i) 0.1; (ii) 0.4.
> In each case a directivity factor of 4 is applicable since the machine is assumed to stand on the floor near to a wall.

Case a (i): Sound pressure level is given by:

$$L_p = L_w + 10 \log_{10}\left(\frac{Q}{4\pi r^2} + \frac{4}{R_c}\right), \text{ where } R_c = \frac{S\bar{\alpha}}{1-\bar{\alpha}}$$

It is easily shown that $S = 1350$ m², hence:

$$R_c = \frac{1350 \times 0.1}{1 - 0.1} = 150$$

and $L_p = 90 + 10 \log_{10}\left(\frac{4}{4\pi \times 5^2} + \frac{4}{150}\right)$

$= 90 + 10 \log_{10}(0.0127 + 0.0267)$
$= 90 + 10 \log_{10}(0.0394) = 90 - 14 =$ **76 dB**

Case a (ii):

$$R_c = \frac{1350 \times 0.4}{1 - 0.4} = 900$$

and $L_p = 90 + 10 \log_{10} \left(\dfrac{4}{4\pi \times 5^2} + \dfrac{4}{900} \right)$

$= 90 + 10 \log_{10} (0.0172) = 72.3$ dB

Case b (i):

$R_c = 150$

and $L_p = 90 + 10 \log_{10} \left(\dfrac{4}{4\pi \times 20^2} + \dfrac{4}{150} \right) = 74.4$ dB

Case b (ii):

$R_c = 900$

and $L_p = 90 + 10 \log_{10} \left(\dfrac{4}{4\pi \times 20^2} + \dfrac{4}{900} \right) = 67.2$ dB

From studying the above results it will be seen that the average absorption coefficient has more effect upon the sound level at greater distances from the source. This is to be expected since near to the source the direct sound is the most important whereas further away from the source the reverberant field becomes important.

SOUND ABSORBING MATERIALS

The sound energy which is absorbed is generally converted into heat. Sound absorbing materials may be broadly classified into three types:
(i) dissipative or porous absorbers,
(ii) panel or membrane absorbers;
(iii) resonator absorbers.

Porous absorbers are typically open textured materials such as glass fibre, mineral wool, open cell plastic foams and acoustic tile materials. The absorption coefficient increases with frequency as shown in *Fig 2*. The absorption coefficient at low frequencies is improved as the thickness of the material is increased.

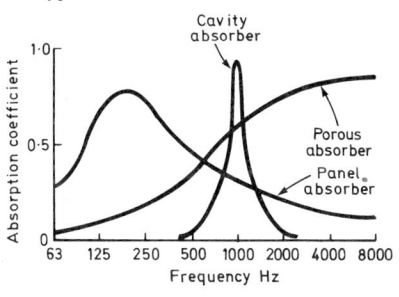

Fig 2 Variation of absorption coefficient with frequency

Panel absorbers absorb energy by the vibrations of the panel. The variation of absorption coefficient with frequency is shown in *Fig 2*. The frequency at which the maximum absorption occurs is given approximately by:

$f_r = \dfrac{60}{\sqrt{md}}$ where f_r = frequency of maximum absorption,
m = mass per unit area of the panel in kg/m^2;
and d = depth of the airspace in metres.

Panel absorbers may occur fortuitously as in the case of a suspended ceiling. If the suspended ceiling consists of acoustic tiles then the overall absorption properties will be a combination of those of a porous material and those of a panel absorber.

Cavity or Helmholtz resonators are containers with a small open neck and absorb sound by resonance of the air within the cavity. They provide a high absorption coefficient over a narrow bond of frequencies as shown in *Fig 2*. For a resonator of the type shown in *Fig 3* the maximum absorption occurs at a frequency given by:

$$f_r = \frac{vr}{2\pi} \sqrt{\frac{2\pi}{(2\ell + \pi r) V}}$$

where v is the velocity of sound in air and l; r and V are the dimensions shown in *Fig 3*.

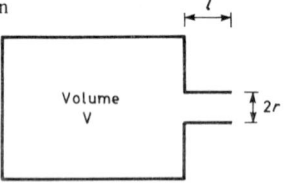

Fig 3 Cavity resonator

Resonators are useful for controlling sound absorption at particular frequencies. Resonators will occur where perforated panels are mounted over an absorbing material; the variation of absorption with frequency is not as peaked as with the true resonator.

> *Problem 2* A suspended ceiling consists of panels having a mass per unit area of 7 kg/m². The depth of the airspace is 0.3 m. Deduce the frequency at which the maximum absorption will occur when the ceiling is considered as a panel absorber.

The required frequency is given by:

$$f_r = \frac{60}{\sqrt{md}} = \frac{60}{\sqrt{7 \times 0.3}} = 41 \text{ Hz}$$

REVERBERATION TIME AND SABINE'S FORMULA

When a sound source in a room is stopped, a certain time is required for the sound energy to die away by absorption at the room surfaces. The reverberation time is the time taken for the sound pressure level to decay by 60 dB. For rooms with hard surfaces having low absorption coefficients this time may be several seconds giving a room which is noisy and has poor acoustics. The reverberation time is one important measure of room acoustics. Methods of prediction of reverberation time are thus important in acoustic design.

The simplest formula for reverberation time is that due to W.C. Sabine, which is:

$$T = \frac{0.16V}{A}$$

where T = reverberation time,
V = volume of room and,
$A = \Sigma \alpha_i S_i$,

α_i being the absorption coefficient of each individual material having an area S_i.

> *Problem 3* The details of a lecture room in a college are shown below. The volume of the room is 245 m³. Calculate the reverbration time at 500 Hz when the room is occupied by 20 people each of whom contribute 0.4 m² of absorption.
>
> | Item | Area (m²) | Absorption coefficient at 500 Hz |
> |---|---|---|
> | Floor | 49 | 0.05 |
> | Walls | 120 | 0.02 |
> | Windows | 20 | 0.10 |
> | Ceiling tiles | 49 | 0.60 |
>
> The audience may be assumed to reduce the floor absorption by 40%.

The calculation of the total absorption is best set out in tabular form

Item	Area, S	Absorption coefficient α	αS
Floor	29.4	0.05	1.47
Walls	120	0.02	2.4
Windows	20	0.10	2.0
Ceiling tiles	49	0.60	29.4
Audience	20 people	0.4 per person	8
		Total ΣαS =	43.27

Note that the floor area has been reduced by 40% since the audience cover this amount of the floor. The reverberation time can now be found using Sabine's formula:

$$T = \frac{0.16V}{A} = \frac{0.16 \times 245}{43.27} = \mathbf{0.9 \text{ s}}$$

Problem 4 A workshop is 30 m by 15 m on plan and 5 m high. The reverberation time was measured and found to be 3 s. Calculate the average absorption coefficient of the room surfaces. Determine the additional absorption required to reduce the reverberation time to 0.9 s

The volume of the room is 2250 m³ and its total surface area is 1350 m². By rearranging Sabine's formula the total absorption, A, is given by:

$$A = \frac{0.16V}{T} = \frac{0.16 \times 2250}{3} = 120 \text{ m}^2$$

Since the area of the room surfaces is 1350 m² the average absorption coefficient $\bar{\alpha}$ is:

$$\bar{\alpha} = \frac{120}{1350} = \mathbf{0.089}$$

The total absorption required to give a reverberation time of 0.9 seconds is found as:

$$A = \frac{0.16V}{T} = \frac{0.16 \times 2250}{0.9} = 400 \text{ m}^2$$

The existing absorption is 120 m² so that:
 additional absorption = 400 − 120 = **280 m²**.

Optimum reverberation time

For many applications there exist optimum reverberation times which depend on the use to which the room is to be put and also the volume of the room. Since absorption coefficients of materials change with frequency the optimum reverberation times are usually given for a frequency of 500 Hz. Some typical optimum reverberation times are shown in *Fig 4*.

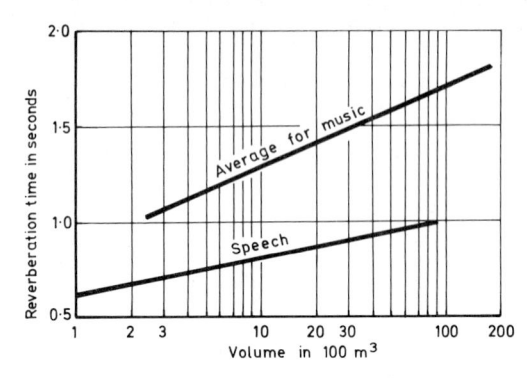

Fig 4 Optimum reverberation times

Eyring's Formula for reverberation time

If the average absorption coefficient exceeds about 0.2 the formula due to Eyring is more accurate. This formula is:

$$T = \frac{0.16V}{-2.30 \, S \log_{10}(1-\bar{\alpha})}$$

At high frequencies and for large rooms the absorption of sound by the air becomes important. The air absorption is only significant for frequencies over 2000 Hz and is chiefly dependant on the relative humidity of the air. The Eyring formula can then be written as

$$T = \frac{0.16V}{-2.30 \, S \log_{10}(1-\bar{\alpha}) + 4mV}$$

where m is the absorption coefficient of the air, some typical values of which are given in *Table 1*

TABLE 1

Relative humidity %	absorption coefficient m	
	2000 Hz	4000 Hz
20	0.006	0.016
50	0.003	0.008
70	0.002	0.006

Problem 5 A large room has a volume of 11300 m³ and a surface area of 4180 m². The average absorption coefficient of the room surfaces is 0.15. Determine the reverberation time:
(i) making no allowance for air absorption;
(ii) assuming that the absorption coefficient of the air is 0.004 per metre.

(i) Using the Eyring formula:

$$T = \frac{0.16 V}{-2.30 \, S \log_{10}(1-\bar{\alpha})}$$

$$= \frac{0.16 \times 11300}{-2.30 \times 4180 \log_{10}(1-0.15)} = 2.7 \text{ s}$$

(ii) Including the air absorption and using the equation stated:

$$T = \frac{0.16 \times 11300}{-2.30 \times 4180 \log_{10}(1-0.15) + 4 \times 0.004 \times 11300}$$
$$= 2.1 \text{ s}$$

DESIGN OF ROOM SHAPE

It is not possible in a book of this size to consider in detail the acoustic design of rooms for all types of use. The following ideas will however establish some of the important points to be considered.

(i) The volume of the room:

The maximum room volume depends upon the power of the source; for instance for an average speaker the maximum room volume would be 3000 m³ whereas for a symphony orchestra the maximum volume would be 20000 m³.

For most public rooms the majority of sound absorption is contributed by the audience and it is thus possible to arrive at an optimum volume per person in order to achieve an approximately correct reverberation time. For speech a volume per person of 3 to 5 m³ is suitable and for concert halls a volume per person of 7 to 9 m³ would be preferable.

(ii) Form

Sound paths for all the audience seats must be as short and as direct as possible. If a rectangular room is used the above requirement would at first sight lead to the use of a square room. This is unsatisfactory for two reasons. Firstly, most sources, particularly the human voice are directional, most of the power being emitted forwards. Seats to the side of a speaker will receive little power. However the length of a rectangular room should not exceed twice its width. Secondly in any room standing waves are created. A standing wave occurs when the distance between two surfaces is a multiple of half the wavelength of the sound. In a room a large number of standing waves with different frequencies will occur; these are referred to as the room modes.

When the source of sound is stopped each of these standing waves will decay by absorption at the room surfaces. For good acoustics each standing wave should decay at approximately the same rate. If the room dimensions are in simple ratios then many modes having the same frequency will occur. Thus this frequency will contain more energy and will decay less rapidly than other modes. One frequency will thus persist longer than other frequencies. In conclusion therefore the room dimensions of a rectangular room should not be in simple ratios such as 1:2:4.

The requirement that the sound path should be as short as possible leads to the use of trapezoidal or fan shaped rooms. For large audiences the use of balconies is helpful in achieving short sound paths. Sound paths must be unobstructed and not pass through or just over the heads of the audience. Tiering of the seats avoids sound being absorbed by the audience.

Large concave surfaces which cause focussing of the sound should be avoided. The focussing effect, shown in *Fig 5*, results in an uneven energy distribution and thus large variations in intensity throughout the room.

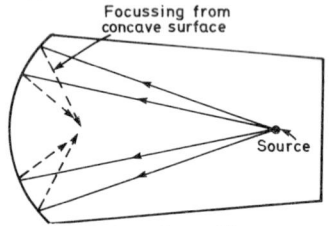

Fig 5 Focusing of sound by a concave surface

Fig 6 Directed and reflected paths from source to receiver

In a room the sound reaches the listener by many paths due to reflections from the surfaces, as shown in *Fig 6*. If a reflected sound is heard more than 35–45 ms later than the direct sound then it will be perceived as an echo. This implies that the path length difference must not exceed 12–15 m, the shorter value being applicable for speech. This problem is more likely to occur in large rooms. Hard parallel surfaces may create flutter echoes with impulsive sounds.

(iii) Reverberation time

It has already been seen that there is an optimum reverberation time for a particular room and usage. In many rooms, particularly those used for speech, additional absorption will be necessary.

Considerable care is necessary in selecting the position for this absorption otherwise a very non-uniform sound field will result and useful reflections may well be lost. In a lecture room, for instance, placing acoustic tiles over the complete ceiling will reduce the energy reflected by the ceiling towards the listeners at the back of the room leading to a loss of intelligibility. It would thus be better to place this absorption on the front third of the ceiling and to spread the remainder on the sides walls.

As a general principle it is better to spread the absorption in patches rather than concentrate it on one room surface and absorption placed near the corners of the room is more effective than elsewhere.

For a more detailed treatment of the acoustics of auditoria the reader is referred to specialist books on this subject.

EXERCISES (answers on page 185)

In the following questions, where options are given select the correct option or options.

1. A workshop is 20 m by 20 m on plan and is 5 m high. A machine of sound power level 100 dB stands on the floor against the middle of one wall. The appropriate directivity factor in calculating the sound pressure level is:
   (a) 1; (b) 2, (c) 4; (d) 8.

2. For the workshop and machine in *Exercise 1*, what is the appropriate directivity factor if the machine is moved into a corner of the workshop?
   (a) 1; (b) 2, (c) 4; (d) 8.

3. For the machine and workshop in exercise 1, calculate the sound pressure level for distances at 2 m intervals from 2 m to 20 m away from the machine. Assume that the average absorption coefficient is 0.2. Illustrate the results graphically.

4. The absorption coefficient of a porous material:
   (a) decreases as the frequency increases;
   (b) increases as the frequency increases;
   (c) is constant for all frequencies,
   (d) is increased at low frequencies by increasing the thickness.

5. A suspended ceiling has a mass per unit area of 8 kg/m^2. The depth of the airspace is 0.2 m. If this ceiling is considered as a panel absorber estimate the frequency at which the absorption coefficient will be a maximum.

6. A lecture room, with a maximum seating capacity of 200 people, has a volume of 800 m^3. The surfaces of the room and their absorption coefficients are given below. Assuming that each person contributes 0.4 m^2 of absorption and each empty seat contributes 0.15 m^2 of absorption, calculate by means of Sabine's formula the reverberation time of the room when
   (i) 50 people are present; (ii) 150 people are present.
   Assume in each case that the audience and seats cover 60% of the floor.

Item	Area (m^2)	Absorption coefficient at 500 Hz
Floor	160	0.05
Walls	195	0.02
Windows	65	0.1
Ceiling	160	0.2

7. A room of volume 500 m^3 was found to have a reverberation time of 2.0 s. Calculate the extra absorption required to reduce the reverberation time to 0.9 s, assume Sabine's formula to be applicable.

8. A recording studio is 5 m by 5 m by 3.2 m high and is required to have a reverberation time of 0.35 s. Use the Eyring formula to calculate the average absorption coefficient of the room surfaces.

9. In designing a room for speech an optimum volume per person would be:
   (1) 1 to 2 m^3; (b) 2 to 3 m^3; (c) 3 to 5 m^3; (d) no restriction exists.

10. In designing a rectangular room for speech the best shape would be:
    (a) square in plan;
    (b) rectangular with length greater than twice the width;
    (c) rectangular with length less than twice the width;
    (d) rectangular with the room dimensions in the ratio 9:3:1 for length:breadth:height.

11 Additional absorption is required in a lecture room.
   The best position for this absorption would be:
   (a) entirely on the ceiling;
   (b) on the front third of the ceiling and the remainder on the walls;
   (c) on one side wall only;
   (d) on any surface but not in the corners of the room.

12 At a certain position in a hall an echo of the sound is heard just after the direct sound reaches the listener. This may be caused by:
   (a) a reflected sound arriving about 40 milliseconds after the direct sound;
   (b) a sound being reflected from a wall close to the listener;
   (c) the sound being reflected from a wall about 7 m behind the listener;
   (d) too short a reverberation time.

7 Sound insulation

AIRBORNE SOUND INSULATION

Consider a partition which separates a sound source in air on one side of it and a receiver in air, often the human ear, on the other side. Airborne sound insulation is concerned with the ability of the partition to reduce the sound pressure level between the source and receiving sides.

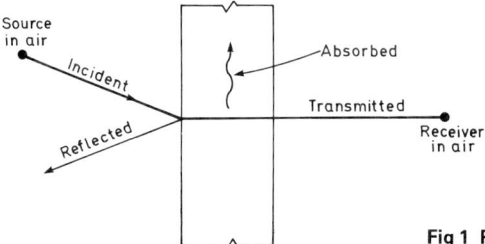

Fig 1 Reflection, absorption and transmission of a sound wave

Fig 1 shows the energy incident upon a partition some of this is reflected, some absorbed by the partition and converted to heat and the rest is transmitted to the receiver. The transmission coefficient, denoted by τ, is defined as

$$\tau = \frac{\text{transmitted energy}}{\text{incident energy}}$$

It is useful to express the sound insulation in the decibel scale and this is termed the sound reduction index, denoted by R and is defined as

$$R = 10 \log_{10} \left(\frac{1}{\tau}\right)$$

The sound reduction index is of fundamental importance in predicting the acoustic properties of a partition.

> *Problem 1* The sound reduction index of a partition is 40 dB. Determine the transmission coefficient.

Using the above formula:

$$40 = 10 \log_{10}\left(\frac{1}{\tau}\right)$$

$$\log_{10}\left(\frac{1}{\tau}\right) = 4$$

Hence, by taking antilogarithms:

$$\frac{1}{\tau} = 10\,000 \text{ giving } \tau = \mathbf{0.0001}$$

This shows that only one ten-thousandth of the incident energy is transmitted.

SOUND REDUCTION INDEX OF SINGLE LEAF PARTITIONS

The variation of sound reduction index of a partition with frequency is shown in *Fig 2*. The various regions of the graph will now be considered.

At low frequencies the sound reduction index depends upon the stiffness of the partition. A stiff panel has a natural frequency of vibration. If a sound wave having this frequency strikes the panel it will resonate and the sound reduction index will be considerably less. All panels have a number of resonant frequencies which depend upon the dimensions of the panel and its edge fixing. The resonant frequency will be lower for panels of low stiffness and it would thus appear to be desirable to reduce the stiffness as much as possible so that the resonant frequency is below the audio range. There are of course practical limitations to this.

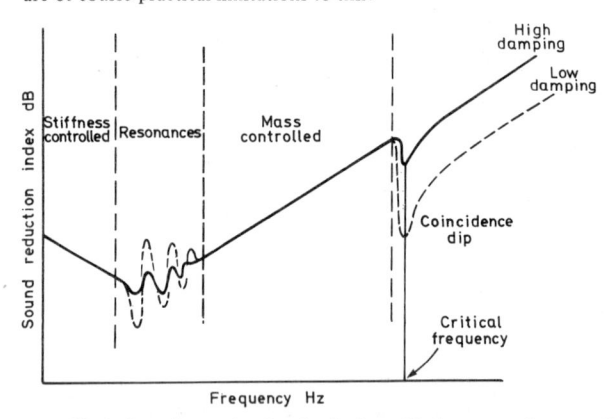

Fig 2 Variation of sound reduction index with frequency for a partition

In the mass controlled region the sound reduction index depends upon the mass of the partition; the greater the mass the greater the sound reduction index. As will be seen in *Fig 2* the sound reduction index also increases with frequency. The two following simple rules will provide a reasonable estimate of the sound reduction index in the mass controlled region.

(i) The average sound reduction index is related to the mass by:

$R_{av} = 10 + 14.5 \log_{10} m$ where m is the mass per unit area in kg/m^2.

This average figure will adequately represent the performance at a frequency of 500 Hz.

(ii) The sound reduction index will increase by approximately 5 dB for each doubling of frequency.

> *Problem 2* Predict the likely sound reduction index at 500 Hz for a brick wall, plastered on both sides, which has a mass per unit area of 250 kg/m². Suggest the likely sound reduction index at the following frequencies: 63, 125, 250, 1000, 2000 Hz.

Using the equation:
$$R_{av} = 10 + 14.5 \log_{10} m$$
$$= 10 + 14.5 \log_{10} 250$$
$$= 45 \text{ dB which is applicable at 500 Hz}$$

The sound reduction index will increase by 5 dB for each doubling of the frequency and will decrease by 5 dB each time the frequency is halved. The following table shows the expected performance:

Frequency (Hz)	63	125	250	500	1000	2000
Sound reduction index (dB)	30	35	40	45	50	55

Referring again to *Fig 2* the decrease in sound reduction index at the coincidence dip is due to the stiffness of the partition. If a panel is vibrated in a direction at right angles to its surface a bending wave will travel along the panel. Similar waves can be seen travelling in a long spring if one end is oscillated at right angles to the length of the spring.

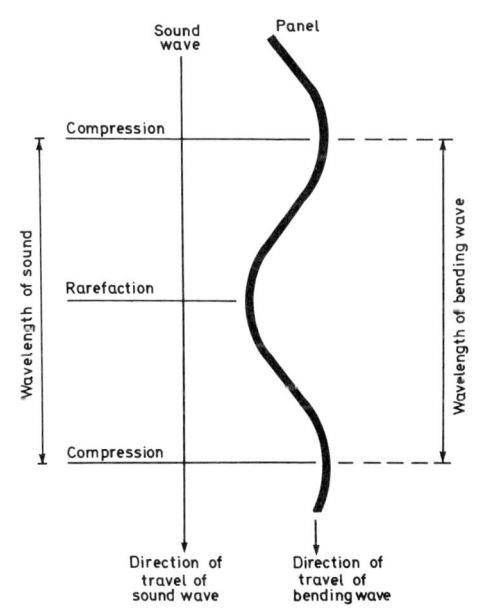

Fig 3 Coincidence effect with sound wave parallel to panel

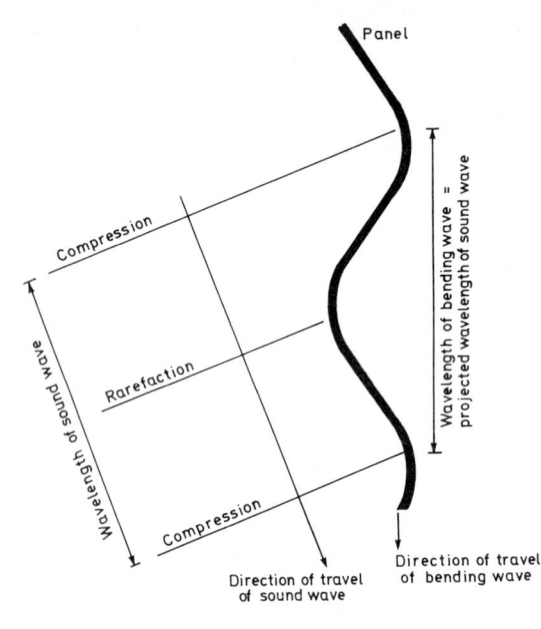

Fig 4 Coincidence effect with sound wave obliquely incident to panel

At suitable frequencies and angles of incidence the compressions and rarefactions in a sound wave travelling near the panel will reinforce the displacement of the panel. *Fig 3* shows the simplest case where the sound wave is travelling parallel to the panel. Since the sound wave is causing the panel to vibrate the sound reduction index will be reduced. In this case the wavelength of the bending wave and the wavelength of the sound wave are equal and the frequency of the sound wave at which this occurs is called the critical frequency.

This is not a single frequency effect since by considering sound waves which are not parallel to the panel it will be seen, in *Fig 4*, that other sound wavelengths can cause coincidence. It will be noted that the projected wavelength of the sound in air is now equal to the wavelength of the bending wave. The decrease in sound reduction index depends upon the stiffness and damping of the material.

An estimate of the critical frequency can be found for some materials by using the data in the table below:

Material	Critical frequency × mass per unit area (Hz × kg/m^2)
Steel	100 000
Brick	42 000
Glass	35 000
Plasterboard	32 000
Plywood	15 000

Problem 3 Estimate the critical frequency for 6 mm glass having a mass per unit area of 15 kg/m².

From the above table it will be seen that:
critical frequency × mass per unit area = 35 000

$$\text{critical frequency} = \frac{35\,000}{\text{mass per unit area}}$$
$$= \frac{35\,000}{15} = 2300 \text{ Hz}$$

As glass possesses little damping the loss in sound reduction index could be considerable.

Composite partitions

Many partitions consist of different constructions of different sound reduction indices, for example walls containing windows and doors.

It is often necessary to assess the overall sound reduction index of such a partition; this can be calculated using the following procedure:

(i) For each of the constructions, having areas $S_1, S_2, S_3 \ldots$, determine the transmission coefficients, $\tau_1, \tau_2, \tau_3 \ldots$, from the sound reduction indices using the method of *Problem 1* in this chapter.

(ii) Find the average transmission coefficient, τ_{av}, using:

$$\tau_{av} = \frac{S_1 \tau_1 + S_2 \tau_2 + S_3 \tau_3 + \ldots}{S_1 + S_2 + S_3 + \ldots}$$

(iii) Find the average sound reduction index, R_{av}.

$$R_{av} = 10 \log_{10} \left(\frac{1}{\tau_{av}}\right)$$

Problem 4 A partition consists of 12 m² of cavity brick wall, 4 m² of single glazed window and 2 m² of hardwood door. The sound reduction indices at 1000 Hz are: 55 dB for the brickwork, 30 dB for the window and 25 dB for the door.

Calculate the average sound reduction index for the partition.

For the brickwork, using the definition of sound reduction index:

$$R = 10 \log_{10} \left(\frac{1}{\tau}\right)$$
$$55 = 10 \log_{10} \left(\frac{1}{\tau_1}\right)$$
$$\log_{10} \left(\frac{1}{\tau_1}\right) = 5.5$$
$$\tau_1 = 3.162 \times 10^{-6}$$

Similarly the transmission coefficient for the window and door are:
$\tau_2 = 1 \times 10^{-3}$ for the window
$\tau_3 = 3.162 \times 10^{-3}$ for the door.

The average sound transmission coefficient is given by:

$$\tau_{av} = \frac{S_1\tau_1 + S_2\tau_2 + S_3\tau_3}{S_1 + S_2 + S_3}$$

$$= \frac{12 \times 3.162 \times 10^{-6} + 4 \times 1 \times 10^{-3} + 2 \times 3.162 \times 10^{-3}}{12 + 4 + 2}$$

$$= 5.757 \times 10^{-4}$$

The average sound reduction index is found as follows:

$$R_{av} = 10 \log_{10}\left(\frac{1}{\tau_{av}}\right)$$

$$= 10 \log_{10}\left(\frac{1}{5.757 \times 10^{-4}}\right) = \mathbf{32.4\ dB}$$

It will be seen that the average sound reduction index is much nearer to that of the door and window than that of the brickwork. Small areas of low sound reduction index will seriously reduce the sound reduction index of an otherwise good partition; this applies particularly to any holes, cracks or air gaps.

Problem 5 A cavity brick wall has a sound reduction index of 55 dB and an area of 12 m². A window is to be built in this wall, the sound reduction index of the window being 30 dB. Find the area of the window which can be built if the average sound reduction index is to be 35 dB.

From *Problem 4* the transmission coefficients are: 3.162×10^{-6} for the brickwork and 1×10^{-3} for the window. The average transmission coefficient is given by:

$$35 = 10 \log_{10}\left(\frac{1}{\tau_{av}}\right)$$

$$\tau_{av} = 3.162 \times 10^{-4}$$

Let x be the area of the window, then the area of the brickwork is $12 - x$. Applying the formula for the average transmission coefficient:

$$3.162 \times 10^{-4} = \frac{3.162 \times 10^{-6}(12 - x) + 1 \times 10^{-3}x}{12}$$

$$3.162 \times 10^{-6}(12 - x) + 1 \times 10^{-3}x = 12 \times 3.162 \times 10^{-4}$$

$$3.794 \times 10^{-5} - 3.162 \times 10^{-6}x + 1 \times 10^{-3}x = 3.794 \times 10^{-3}$$

$$9.9684 \times 10^{-4}x = 3.7560 \times 10^{-3}$$

$$x = 3.8\ m^2$$

Thus the maximum area of the window will be **3.8 m²**.

Problem 6 A cavity brick wall has a sound reduction index of 55 dB and an area of 12 m². A window of area 5 m² is to be built in this wall and the average sound reduction index is to be 43 dB. Calculate the necessary sound reduction index for the window.

As in *Problem 4*, the transmission coefficient for the brickwork is 3.162×10^{-6}. The average transmission coefficient is given by

$$43 = 10 \log_{10}\left(\frac{1}{\tau_{av}}\right)$$

$$\tau_{av} = 5.012 \times 10^{-5}$$

Applying the formula for the average transmission coefficient, in which the transmission coefficient of the window is denoted by τ_2:

$$5.012 \times 10^{-5} = \frac{7 \times 3.162 \times 10^{-6} + 5\tau_2}{12}$$

$$5\tau_2 = 12 \times 5.012 \times 10^{-5} - 7 \times 3.162 \times 10^{-6}$$

$$\tau_2 = 1.158 \times 10^{-4}$$

The sound reduction index of the window is:

$$R = 10 \log_{10}\left(\frac{1}{\tau_2}\right)$$

$$= 10 \log_{10}\left(\frac{1}{1.158 \times 10^{-4}}\right) = 39.4 \text{ dB}$$

Flanking transmission

The sound reduction index discussed above is concerned with the sound that is transmitted directly from the source, through the partition, to the receiver. Sound which reaches the receiver by other paths through the construction is termed flanking transmission. *Fig 5* shows some of the possible flanking transmission paths.

In practice the sound insulation achieved by a partition will depend to a great extent on the flanking transmission. The amount of flanking transmission will depend upon the mass of the supporting structures and the fixing of the partition to these structures.

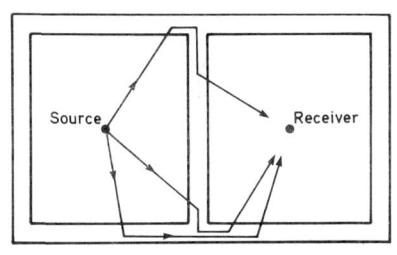

Fig 5 Flanking transmission paths

Many partitions do not realise their full potential as regards sound insulation due to flanking transmission. For example the insulation of many partitions between bedrooms is reduced by a relatively poor ceiling sound insulation allowing the sound to pass through the ceiling into the loft space and thus into another bedroom.

Multiple leaf partitions

The sound reduction index of a partition can, in general, be increased by the use of cavity construction. If a single leaf partition has a sound reduction index of 35 dB and its mass is doubled by doubling its thickness then the sound reduction index is likely to be about 40 dB. If the partition was built as two leafs with a large separation then the sound reduction index would be 35 + 35 = 70 dB, however this is not achieved in practice.

The overall insulation is limited by flanking transmission at the edges and connections between the leaves. As cavities are generally quite narrow the air contained within it acts as a spring and transmits sound from one leaf to the other particularly at low frequencies. The two leaves coupled by the springiness of the air gives rise to resonances which further reduce the sound reduction index.

For ordinary cavity wall construction the average sound reduction index is only 2 or 3 dB better than for a solid wall of the same mass per unit area.

With lightweight partitions cavity construction is more effective since flanking transmission is relatively less important for lower values of sound reduction index. The resonances can be damped by including absorbant material in the cavity. Careful design can reduce the mechanical coupling of the two leaves to a minimum. In some instances it may be practical to include materials having high damping in the construction.

A simple example of the effectiveness of cavity construction is the double glazed window. If this is well sealed, has an airspace of 100 mm or more and the reveals are lined with absorbant material then a sound reduction index approaching twice that of a single glazed window is possible.

MEASUREMENT OF AIRBORNE SOUND INSULATION

The measurement of sound insulation for building elements and between rooms in buildings is covered by BS 2750 (Parts 3 and 4). The sound reduction index for building elements can be measured using laboratory facilities in which the source and receiver rooms are suitably designed so that the flanking transmission is negligible.

A reverberant sound field is created in the source room having a sound pressure level L_1. The sound pressure level, L_2, in the receiving room, suitably averaged for a number of microphone positions, is obtained at each of the one-third octave frequencies from 100 Hz to 3150 Hz. The sound pressure level in the receiving room depends upon the area of the partition and also the total absorption of the materials of the surfaces of the receiving room. In order to find the sound reduction index these factors must be allowed for.

Thus the sound reduction index, R, is given by:

$R = L_1 - L_2 + 10 \log_{10} S - 10 \log_{10} A$

where S is the area of the partition and A is the total absorption given by $\Sigma \alpha_i S_i$.

Similar tests can be conducted for rooms in buildings but in this case flanking transmission will be present. The apparent sound reduction index, R', may be calculated in a similar manner to that used in laboratory tests. For some applications it is useful to correct the level difference between the source and receiving rooms to standardised conditions in the receiving room. For this purpose the level difference, D, is defined as: $D = L_1 - L_2$. The standardised level difference D_{nT} is defined as:

$$D_{nT} = D + 10 \log_{10} \frac{T}{T_o}$$

where T is the reverberation time of the receiving room and T_o is a reference reverberation time which is taken as 0.5 s for domestic dwellings.

PREDICTION OF SOUND LEVELS THROUGH PARTITIONS

The following formulae are a useful guide for the prediction of sound levels where the sound field in the room is reverberant. There are three cases to consider.
(i) Room to Room:
$L_2 = L_1 - R + 10 \log_{10} S - 10 \log_{10} A$
(ii) Room to outside:
$L_2 = L_1 - R + 10 \log_{10} S - 20 \log_{10} r - 14$
(iii) Outside to inside
$L_2 = L_1 - R + 10 \log_{10} S - 10 \log_{10} A + 6$
where L_2 = sound pressure level on the receiving side;
L_1 = sound pressure level on the source side;

R = appropriate sound reduction index of the partition;
S = area of partition;
A = total absorption of receiving room = $\Sigma \alpha_i S_i$;
r = distance from the partition.

Problem 7 An office 10 m by 12 m by 3 m high is separated on its longer side from a workshop by a solid partition whose sound reduction index is 34 dB at 250 Hz. The details of the office surfaces and absorption coefficients at 250 Hz are as follows:

Item	Area (m^2)	Absorption coefficients, α
Acoustic tiles on ceiling	120	0.35
Wall (excluding windows)	100	0.05
Windows	32	0.25
Floor (carpeted)	120	0.08
People	15 persons	0.04 per person

If the reverberant sound pressure level in the workshop is 85 dB at 250 Hz, estimate the reverberant sound pressure level in the office at this frequency.

The first step is to find the total absorption of the office:
$A = \Sigma \alpha_i S_i$
  $= (120 \times 0.35) + (100 \times 0.05) + (32 \times 0.25) + (120 \times 0.08) + (15 \times 0.4)$
  $= 70.6 \text{ m}^2$

It should be noted that the area of the partition wall is $12 \times 3 = 36 \text{ m}^2$. Using the formula quoted above:
$L_2 = L_1 - R + 10 \log_{10} S - 10 \log_{10} A$
   $= 85 - 34 + 10 \log_{10} 36 - 10 \log_{10} 70.6$
   $= 85 - 34 + 15.6 - 18.5 = \textbf{48 dB}.$

This only applies at 250 Hz and for a full analysis of the acoustic behaviour of the partition the analysis should be carried out at each of the third octave frequencies from 100 Hz to 3150 Hz.

Problem 8 The external wall of a machine shop consists of 30 m^2 of brick wall having an average sound reduction index of 45 dB and 15 m^2 of glazing having an average sound reduction index of 22 dB. The reverberant sound pressure level in the machine shop is 85 dB. A domestic dwelling stands 15 m from the machine shop wall. Calculate
(i) the sound reduction index of the external wall;
(ii) the sound pressure level at the outside of the domestic dwelling.

(i) The sound reduction index of the external wall is found using the method of *Problem 4* in this chapter.
  For the brickwork:
$R = 10 \log_{10} \left(\frac{1}{\tau}\right)$
$45 = 10 \log_{10} \left(\frac{1}{\tau_1}\right)$
$\tau_1 = 3.162 \times 10^{-5}$

Similarly for the window the transmission coefficient $\tau_2 = 6.31 \times 10^{-3}$. The average transmission coefficient is:
$$\tau_{av} = \frac{30 \times 3.162 \times 10^{-5} + 15 \times 6.31 \times 10^{-3}}{30 + 15} = 2.12 \times 10^{-3}$$
The sound reduction index is found by:
$$R_{av} = 10 \log_{10}\left(\frac{1}{\tau_{av}}\right)$$
$$= 10 \log_{10}\left(\frac{1}{2.12 \times 10^{-3}}\right) = \mathbf{26.7 \; dB}$$

(ii) The sound pressure level at the dwelling may be estimated using:
$L_2 = L_1 - R + 10 \log_{10} S - 20 \log_{10} r - 14$
$L_2 = 85 - 26.7 + 10 \log_{10}(30 + 15) - 20 \log_{10} 15 - 14$
$= 85 - 26.7 + 16.5 - 23.5 - 14 = \mathbf{37 \; dB}.$

IMPACT SOUND INSULATION

In this case the noise source is impact upon the structure, a simple example is that of footsteps on a floor. A floor with a good airborne sound reduction index may provide very little impact sound insulation; a solid concrete floor is a good example.

Although the vibration of the floor due to impact is reduced if the mass of the floor is increased, by far the best way of improving the impact sound insulation is to introduce a resilient layer to absorb the energy of the impact. This may be achieved by the use of resilient floor coverings or by the use of a floating floor construction.

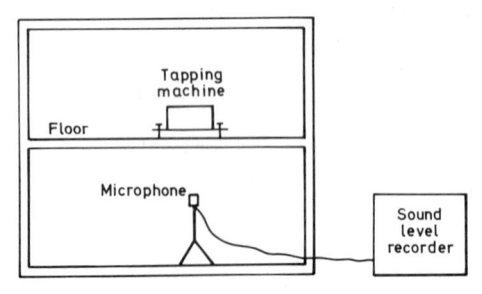

MEASUREMENT OF IMPACT SOUND INSULATION OF FLOORS

This measurement is fully described in BS 2750 (Parts 6 and 7) and only a very brief outline will be considered here. A standard tapping machine is placed upon the floor under test. This machine has five metal hammers, each having a mass of 0.5 kg and falling 40 mm. The hammers are driven by a motor so that 10 impacts per second are created. The sound pressure level in the receiving room (see *Fig 6*) suitably averaged over a number of microphone positions, is obtained at each of the one-third octave frequencies from 100 Hz to 3150 Hz.

As with airborne sound insulation the sound pressure level in the receiving room depends upon the absorption in the receiving room. It is thus necessary to correct

Fig 6 Measurement of impact sound insulation

for this; two methods are possible. In the first method the normalised impact sound pressure, L_n, is defined as

$$L_n = L_1 + 10 \log_{10} \frac{A}{A_o}$$

where L_1 is the sound pressure level in the receiving room,

A is the absorption of the receiving room;

A_o is a reference amount of absorption usually taken as 10 m^2.

As an alternative the standardised impact sound pressure level, L'_{nT}, may be used for floors between rooms in buildings:

$$L'_{nT} = L_1 - 10 \log_{10} \frac{T}{T_o}$$

where T is the reverberation time of the receiving room,

T_o is a reference reverberation time which is taken as 0.5 s for domestic dwellings.

REQUIREMENTS FOR SOUND INSULATION

Section G of the Building Regulations covers the necessary sound insulation between dwellings. The following is only a brief outline of some aspects of the Regulations and for more detail the reader is advised to consult the original document.

The sound insulation of walls is deemed to satisfy if the sound reduction, as measured and normalised in accordance with the appropriate BS 2750, has values not less than those in *Table 1*.

TABLE 1 SOUND REDUCTION: WALLS

Frequency (Hz)	100	125	160	200	250	315	400	500
Sound reduction (dB)	40	41	43	44	45	47	48	49
Frequency (Hz)	630	800	1000	1250	1600	2000	2500	3150
Sound reduction (dB)	51	52	53	55	56	56	56	56

Some values below those in the table are permitted as long as the total of such deviations does not exceed 23 dB.

For floors required to resist the transmission of airborne and impact sound, the sound reduction and normalised impact sound pressure level are shown in *Table 2*. In the case of the sound reduction some lower figures are acceptable as long as the total of such deviations does not exceed 23 dB. For the impact sound pressure level some higher figures are permitted as long as the total of such deviations does not exceed 23 dB.

Schedule 12 of the Regulations gives a number of deemed to satisfy constructions.

> *Problem 9* A floor construction is required to satisfy for both airborne and impact sound transmission. The results given in *Table 3* were obtained when this construction was tested in accordance with BS 2750.
> Determine whether this floor construction will satisfy the requirements of the Building Regulations.

A comparison of the sound reduction of the floor with the requirements for the sound reduction given in *Table 2* shows that in all cases the sound reduction of the floor is

TABLE 2 SOUND REDUCTION: FLOORS

Frequency (Hz)	Sound reduction (dB)	Impact sound pressure level (dB)
100	36	63
125	38	64
160	39	65
200	41	66
250	43	66
315	44	66
400	46	66
500	48	66
630	49	65
800	51	64
1000	53	63
1250	54	61
1600	56	59
2000	56	57
2500	56	55
3150	56	53

TABLE 3 SOUND REDUCTION OF FLOOR IN PROBLEM 9

Frequency (Hz)	Sound reduction (dB)	Impact sound pressure level (dB)
100	38	68
125	38	67
160	40	66
200	42	68
250	43	70
315	45	70
400	47	70
500	48	69
630	51	67
800	52	64
1000	54	61
1250	56	58
1600	59	53
2000	61	50
2500	60	48
3150	63	46

equal to or exceeds the required values and from the airborne sound insulation point of view the floor is satisfactory. The impact sound pressure level for the floor exceeds the required figures for all frequencies from 100 Hz to 630 Hz. It is now necessary to determine whether the total of the deviations exceeds 23 dB. This is illustrated in *Table 4*.

Note that, although the impact sound pressure levels for all frequencies above 800 Hz are below those given in the Regulations one may not offset the good insulation at

TABLE 4

Frequency (Hz)	Sound pressure level for floor (dB)	Sound pressure level in Regulations (dB)	Deviation
100	68	63	5
125	67	64	3
160	66	65	1
200	68	66	2
250	70	66	4
315	70	66	4
400	70	66	4
500	69	66	3
630	67	65	2
800	64	64	0
1000	61	63	—
1250	58	61	—
1600	53	59	—
2000	50	57	—
2500	48	55	—
3150	46	53	—
		Total	28

high frequencies against the poor insulation at low frequencies. The sum total of the adverse deviations is 28 dB which is in excess of the permissible 23 dB and the floor does not satisfy the requirements of the Regulations.

RATING SOUND INSULATION IN BUILDINGS

The reader will by now appreciate that both airborne and impact sound insulation properties are frequency dependent. It is useful to be able to reduce the sound insulation of a building element to a single figure so that comparison of different elements can be made. It has been customary particularly for airborne sound insulation, to find the arithmetic mean of the sixteen values of the sound reduction index obtained at the one-third octave frequencies from 100–3150 Hz. This average has then been used for comparing different constructions but does not take account of the subjective impression of loudness which varies with the frequency of the sound.

BS 5821 gives a method of producing a single-figure rating of the sound insulation performance of a building element by weighting the sound intensity of each frequency according to the subjective impression of loudness. This allows building elements to be compared in accordance with their ability to attenuate sound as perceived by the human ear. Space precludes a full discussion of this British Standard and the reader is referred to BS 5821 for the details.

EXERCISES (answers on page 185)

In the following exercises where options are given select the correct option or options.

1. The sound transmission coefficient is defined as:
   (a) $\dfrac{\text{transmitted energy}}{\text{incident energy}}$
   (b) $\dfrac{\text{absorbed energy}}{\text{incident energy}}$
   (c) $\dfrac{\text{reflected energy}}{\text{transmitted energy}}$
   (d) $\dfrac{\text{incident energy}}{\text{transmitted energy}}$
   (e) $\dfrac{\text{absorbed energy}}{\text{transmitted energy}}$

2. The sound reduction index of a partition is 27 dB. Calculate the sound transmission coefficient.

3. The sound transmission coefficient of a partition is 0.0006. Calculate the sound reduction index.

4. A single leaf partition has a sound reduction index of 28 dB at 500 Hz. The probable sound reduction index at 1000 Hz is:
   (a) 33 dB; (b) 38 dB; (c) 23 dB; (d) 56 dB.

5. A single leaf partition has a mass per unit area of 180 kg/m². Predict the likely sound insulation at the following frequencies:
   63, 125, 250, 500, 1000, 2000 Hz.

6. A single leaf partition is required to have an average sound reduction index of 20 dB. Calculate a suitable mass per unit area for this partition.

7. 6 mm glass has a mass per unit area of 15 kg/m² and a coincidence frequency of 2300 Hz. The coincidence frequency for 3 mm glass will be:
   (a) 2300 Hz; (b) 4600 Hz; (c) 1150 Hz; (d) 5750 Hz.

8. A partition consists of 10 m² of brickwork and 5 m² of window. At a particular frequency the sound reduction indices are 50 dB for the brickwork and 29 dB for the window. The sound reduction index for this partition will be:
   (a) less than 29 dB;   (b) between 29 and 50 dB;
   (c) in excess of 50 dB;   (d) 79 dB.

9. Calculate the overall sound reduction index for a partition having the following specification:
   Brickwork: 14 m² with sound reduction index of 50 dB
   Window: 4.5 m² with sound reduction index of 28 dB
   Door: 2 m² with sound reduction index of 22 dB.

10. The external wall of a lecture room faces a main road and has an area of 30 m². The wall is to be built of cavity brickwork and contain double glazed windows. The sound reduction indices of the brickwork and windows are 55 dB and 40 dB respectively at 1000 Hz. Calculate the permissible area of window if the sound reduction index of the wall is to be 45 dB at 1000 Hz.

11. A large ground floor room with a solid concrete floor was partitioned into two seminar rooms with a blockwork partition, plastered on both sides and having a mass per unit area of 130 kg/m². Following complaints concerning the sound insulation of this partition the average sound reduction index was measured as 25 dB. Which of the following statements and suggestions are reasonable?

(a) this would be the expected sound reduction index for a partition of this mass.
(b) cover the partition with acoustic tiles.
(c) investigate the sealing around large diameter heating pipes which pass through the partition.
(d) investigate possible flanking transmission paths where the partition meets the existing walls and ceiling.

12 A room 5 m by 6 m by 3 m high has a reverberation time of 2 seconds. This room is separated from a workshop by a partition of area 18 m². At a frequency of 250 Hz, the sound reduction index of the partition is 30 dB and the reverberant sound pressure level in the workshop is 75 dB. Calculate:
(a) the absorption of the room using Sabine's formula;
(b) the reverberant sound pressure level in the room.

13 For the situation described in exercise 12 suggest the effect on the reverberant sound pressure level in the room if the room is treated with absorbant material and the reverberation time is reduced to 0.6 s.
(a) it will increase;
(b) it will decrease by more than 10 dB;
(c) it will increase by at least 5 dB;
(d) it will decrease by at least 5 dB.

14 At a particular frequency the sound reduction index of a factory wall is 20 dB, and the reverberant sound pressure level inside the factory is 78 dB. Estimate the sound pressure level at the boundary of the factory premises 20 m away from the wall. The area of the wall of the factory facing this boundary is 30 m².

15 At a particular frequency the sound pressure level due to traffic noise outside a classroom is 71 dB. The external wall of the classroom has an area of 36 m² and a sound reduction index of 28 dB. The details of the classroom are shown in *Table 5*.

TABLE 5

Item	Area (m²)	Absorption coefficient
Ceiling	108	0.2
Walls	96	0.05
Windows	30	0.1
Floor	108	0.08
People	30 people	0.4 per person

Estimate the reverberant sound pressure level in the classroom at this particular frequency.

16 The sound insulation of a party wall was measured and gave the following results:

Frequency	Sound reduction dB
100	38
125	40
160	42
200	43
250	45
315	46
400	48
500	50
630	52
800	53
1000	55
1250	56
1600	58
2000	60
2500	61
3150	63

Considering the requirements of Part G of the Building Regulations for the sound reduction of party walls which of the following statements is correct:

(a) the wall is unsatisfactory since the sound reduction is inadequate at low frequencies,

(b) the wall is unsatisfactory since the sound reduction is inadequate at high frequencies,

(c) the wall is satisfactory since the sum of the adverse deviations is less than 23 dB;

(d) the wall is satisfactory since the good insulation at high frequencies offsets its poor insulation at low frequencies.

8 Hearing and the noise environment

THE HEARING MECHANISM

The hearing mechanism consists of three main divisions, the outer, middle, and the inner ear. *Fig 1(a)* shows a sketch of the hear mechanism and *Fig 1(b)* a schematic representation.

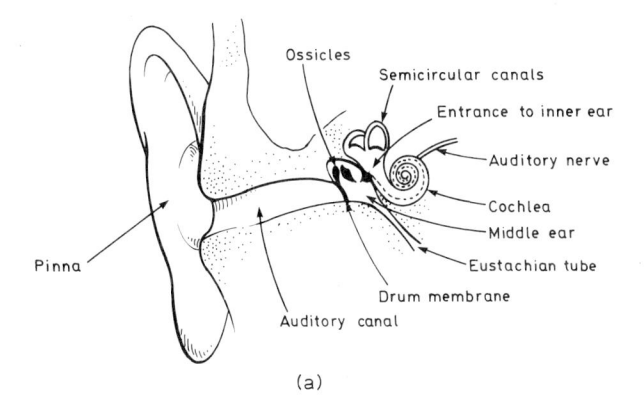

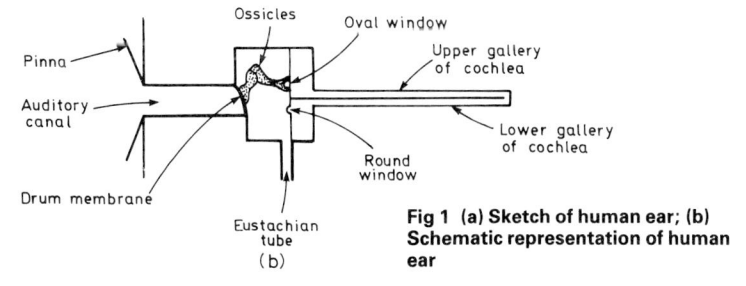

Fig 1 (a) Sketch of human ear; (b) Schematic representation of human ear

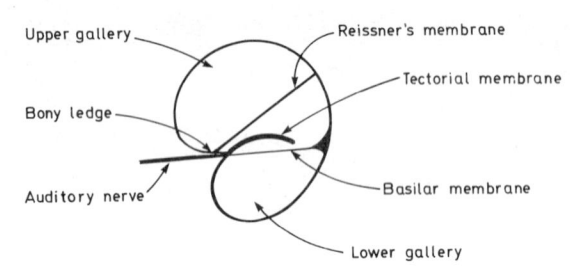

Fig 2 Cross-section of cochlea

The outer ear includes the visible ear or pinna, the auditory canal and the drum membrane or eardrum. The pinna serves as a horn to receive sound energy and leads it into the auditory canal. The auditory canal is an approximately straight tube about 25 mm long and is closed at the inner end by the drum membrane. The resonant frequency of the auditory canal is about 3000 Hz; at this resonant frequency the sound pressure level at the drum is about 10 dB higher than that at the entrance to the auditory canal. Since the resonance curve is quite broad a gain in sound pressure level occurs over the frequency range 2000 to 6000 Hz. This accounts for the increased sensitivity of the human ear over this frequency range.

Inside the drum membrane is the cavity of the middle ear and contains three bones called the ossicles. The ossicles which are called the malleus (hammer), the incus (anvil) and the stapes (stirrup) transmit the vibration of the eardrum to another membrane termed the oval window which connects the middle and inner ear. For high sound intensities the muscles controlling the motion of the ossicles change their tensions so that the motion of the stapes is reduced thus helping to protect the inner ear from damage. The middle ear is connected to the throat by the eustachian tube which opens during swallowing or yawning to allow equilisation between internal and external air pressures.

The inner ear, which is filled with fluid, has three parts: the vestibule or entrance chamber, the semicircular canals and the cochlea. The semicircular canals play no part in the hearing process but provide the sense of balance. The cochlea which is a coiled tube in the shape of a snail shell makes about two and three quarters turns and has a length of about 30 mm.

The tube of the cochlea is divided into two galleries as shown in *Fig 2*. The bony ledge carries the auditory nerve from which the nerve fibres enter the basilar membrane. In the basilar membrane the nerve fibres terminate in minute hair cells which project from its upper surface and touch the tectorial membrane. The vibration of the basilar membrane is detected by these hair cells and thus via the auditory nerves create the sensation of hearing.

HEARING LOSS AND DAMAGE

The sensitivity of the human ear decreases with age, this is called presbycusis. Hearing loss due to presbycusis increases with age and is greatest at high frequencies.

The threshold of hearing can be affected by external noise. If a person is subjected to a fairly intense sound for a short period it will be found that their threshold of hearing has been shited and their ear is less sensitive. After a suitably long rest the threshold of hearing will return to normal. This temporary threshold shift is at its maximum at 4000 Hz irrespective of the frequency of the original sound. If the length of exposure to the noise is increased then the temporary threshold shift will be greater and the duration of the recovery period will increase.

If a person is repeatedly exposed to the noise before the temporary threshold shift has had time to recover a permanent threshold shift will result. As the duration of repeated exposure increases the threshold shift spreads progressively from 4000 Hz down to the lower frequencies and thus interferes with a persons ability to hear speech. This hearing damage induced by noise is additive to the presbycusis effect mentioned above.

The Health and Safety at Work Act requires that people are not exposed to unreasonable noise levels. At present the guidance on noise levels for employed persons is in the Code of Practice for Reducing the Exposure of Employed Persons to Noise produced by the Department of Employment.

CODE OF PRACTICE FOR REDUCING THE EXPOSURE OF EMPLOYED PERSONS TO NOISE

The concept embodied in this code is that an employed person should not receive a noise dose in any one working day in excess of that dose which would occur if the person was exposed to a level of 90 dBA for 8 hours.

For a person exposed to a continuous noise for 8 hours the sound level must not exceed 90 dBA. When the duration of exposure is for a period other than 8 hours or if the sound level is fluctuating an equivalent continuous sound level (L_{eq}) may be found and this should not exceed 90 dBA. The following problems will endeavour to show the concept of an equivalent continuous sound level as the mathematical definition is somewhat complex at first sight.

> *Problem 1* Find the equivalent continuous sound level for a sound of 92 dBA for 2 hours and 86 dBA for 6 hours.

In this case the total duration is 8 hours so that it is necessary to average the sound level over this period. As usual it is not possible to average decibels so that as in the addition and subtraction of decibels the squared pressures are found first (see Chapter 5). For the 92 dBA sound:

$$92 = 10 \log_{10} \frac{p_A^2}{p_o^2}$$

The subscripted p_A is to indicate that the values are obtained using the A-weighting (see Chapter 5). Proceeding in the usual manner:

$$\frac{p_A^2}{p_o^2} = 1.584 \times 10^9$$

For the 86 dBA sound applying the same principles:

$$\frac{p_A^2}{p_o^2} = 3.98 \times 10^8$$

The time average of p_A^2/p_o^2 is then found by multiplying each value of p_A^2/p_o^2 by its appropriate duration, summing the results and finally dividing by the total duration of 8 hours, which gives:

$$\frac{1.584 \times 10^9 \times 2 + 3.98 \times 10^8 \times 6}{8} = 6.945 \times 10^6$$

The value of L_{eq} is now obtained by converting the time average of p_A^2/p_o^2 to decibels in the usual way.
$L_{eq} = 10 \log_{10} 6.945 \times 10^6 =$ **88.4 dBA.**

> *Problem 2* Find the L_{eq} (8 hour) for a person exposed to 95 dBA for 3 hours and 87 dBA for 4 hours during a 7 hour working day.

For the 95 dBA sound: $\dfrac{p_A^2}{p_o^2} = 3.162 \times 10^9$

For the 87 dBA sound: $\dfrac{p_A^2}{p_o^2} = 5.01 \times 10^8$

The time average p_A^2/p_o^2 is now found as before over the 8 hour period even though the working day is only 7 hours:
$$\frac{3.162 \times 10^9 \times 3 + 5.01 \times 10^8 \times 4}{8} = 1.436 \times 10^9$$

The L_{eq} is given by converting this to decibels:
$L_{eq} = 10 \log_{10} 1.436 \times 10^9 =$ **91.6 dBA**

It is hoped that the above problems illustrate the method of averaging used in finding L_{eq} but simple methods would be of practical advantage. In the Code of Practice a nomogram is given which will make this process much simpler. Alternatively the following calculation method can be used:

(i) For each sound level, L, of duration t hours find the fractional exposure, f, using:

$$f = \frac{t}{8} \text{ antilog } [(L - 90)/10]$$

(ii) Find the sum of each of the values of f found in (i). Note that an L_{eq} of 90 dBA for the 8 hour period will give a total f of 1.0. A total f in excess of 1.0 indicates that the L_{eq} exceeds 90 dBA.

(iii) Find $L_{eq} = 10 \log_{10} f + 90$.

The method is basically the same as that used in *Problems 1 and 2* above. It is found that sound levels below 85 dBA do not contribute significantly to the value of L_{eq} and can thus be ignored.

> *Problem 3* Find the L_{eq} (8 hour) for a person exposed to 95 dBA for 3 hours, 87 dBA for 4 hours and 80 dBA for 1 hour.

Following the above stages:
(i) Since the 80 dBA is below the 85 dBA mentioned above it can be ignored in the calculations. For the 95 dBA level:

$f = \dfrac{3}{8}$ antilog $[(95 - 90)/10]$

$= \dfrac{3}{8}$ antilog (0.5)

$= 1.186$

Similarly for the 87 dBA level:

$f = \dfrac{4}{8}$ antilog $[(87 - 90)/10]$

$= \dfrac{4}{8}$ antilog $[-0.3]$

$= 0.251$

(ii) Total $f = 1.186 + 0.251$
       $= 1.437$
(iii) $L_{eq} = 10 \log_{10} 1.437 + 90 =$ **91.6 dBA**
It will be noted that this value exceeds the 90 dBA specified in the code.

> *Problem 4* Find the maximum time of exposure to a sound level of 100 dBA if for the rest of the 8 hour working day the sound level is below 85 dBA.

The period for which the sound level is below 85 dBA can be ignored, thus only the 100 dBA need to be considered. The L_{eq} of 90 dBA will be achieved when the fractional exposure for this sound is 1.0. Let t be the duration of this sound then using:

$f = \dfrac{t}{8}$ antilog $[(L - 90)/10]$

$1.0 = \dfrac{t}{8}$ antilog $[(100 - 90)/10]$

$1.0 = \dfrac{t}{8}$ antilog (1.0)

$1.0 = \dfrac{t}{8} \times 10$

$t = 0.8$

Thus the maximum duration for the 100 dBA sound is **0.8 hours or 48 minutes**.

The above gives but a brief outline of the concepts embodied in the Code of Practice and the reader is advised to read the Code.

NOISE ACCEPTABILITY AND ANNOYANCE

Unfortunately a unique criterion for noise acceptability does not exist. Many people enjoy, and even pay for, high sound pressure levels at discos. The same people will be annoyed, in other circumstances, by much lower sound pressure levels for example the dripping of a water tap in the middle of the night. At the present time a number of criteria are in use and also a number of methods of evaluating the effect of noise which are applied in different situations. The remainder of this chapter will be devoted to a brief introduction to some of these methods.

Speech interforence level (SIL)

The SIL set a maximum level of background noise so that clarity of speech would be preserved. As the intelligibility of speech depends on the frequency spectrum of the background noise the SIL was defined as the arithmetic mean of the sound pressure levels in the octave bands 600–1200 Hz, 1200–2400 Hz and 2400–4800 Hz. The value of the SIL obviously depends on the distance between the speaker and the listener and a few values for normal male speech are given below:

Distance between speaker and listener (m)	0.1	0.5	1.0	2.0	4.0
SIL (dB)	73	61	54	48	42

The SIL alone was found not to be a good guide to the acceptability of the noise level in the room. This led to the development of the Noise Criteria (NC) curves and the Noise Rating (NR) curves. The latter are more widely used in Europe.

Noise rating curves

Fig 3 shows the NR curves which give the maximum sound pressure level at each of the octave bands from 63 to 8000 Hz. *Table 1* gives some recommended NR levels for different environments.

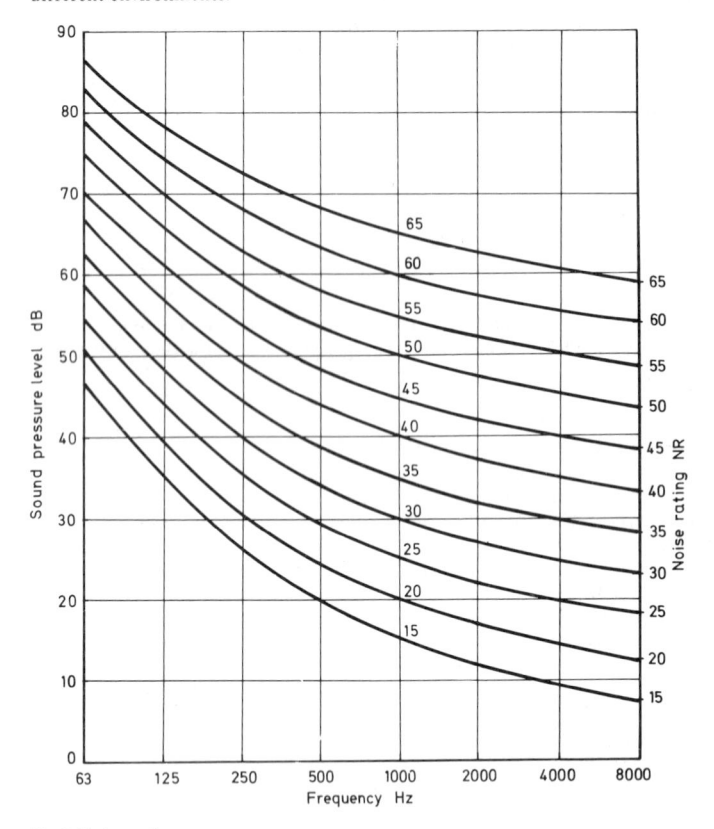

Fig 3 Noise rating curves

TABLE 1 Recommended NR levels

Concert and opera halls	NR 20
Bedrooms in private houses, large conference and lecture rooms	NR 25
Private living rooms, libraries, bedrooms in hotels, operating theatres	NR 30
Public rooms in hotels, middle management and small offices, school classrooms	NR 35
Drawing offices, large restaurants, department stores, recreation rooms	NR 40
Computer rooms, mechanised offices, supermarkets	NR 45

Problem 5 An office 10 m by 12 m by 3 m high is separated on its longer side from a workshop by a partition. The sound pressure levels in the workshop and the reverberation times of the office are given in the table below.

Frequency (Hz)	63	125	250	500	1000	2000	4000	8000
Sound pressure level (dB)	70	79	81	83	80	81	79	75
Reverberation time (sec)	2.30	1.66	0.82	0.42	0.35	0.35	0.37	0.38

Determine (i) the acoustic absorption of the office at each frequency; and (ii) the sound reduction index of the partition at each frequency if the sound level in the office is not to exceed NR 35.

For part (i) the acoustic absorption can be found using Sabine's formula for reverberation time:

$$T = \frac{0.16\,V}{A}$$ which, on rearrangement, gives: $A = \frac{0.16\,V}{T}$

Considering the frequency of 63 Hz at which the reverberation time is 2.30 seconds, the absorption is:

$$A = \frac{0.16 \times 360}{2.30} = 25 \text{ m}^2$$

Similar calculations can be performed at each of the frequencies, the results of which are shown in the table, which also gives value of $10 \log_{10} A$ which will be useful in part (ii)

Frequency (Hz)	63	125	250	500	1000	2000	4000	8000
A m²	25	34.7	70.2	137.1	164.6	164.6	155.7	151.6
$10 \log_{10} A$	14.0	15.4	18.5	21.4	22.2	22.2	21.9	21.8

For part (ii) the reader will recall from Chapter 7 that the sound pressure level room to room through a partition is given by:

$$L_2 = L_1 - R + 10 \log_{10} S - 10 \log_{10} A$$

where in this case L_2 is the sound level in the office which is not to exceed the NR 35 curve at each frequency, L_1 is the given sound pressure level in the workshop, S is the area of the partition which is 36 m² and values of $10 \log_{10} A$ are tabulated above. By rearranging the above equation the sound reduction index, R, is:

$R = L_2 - L_1 + 10 \log_{10} S - 10 \log_{10} A$

The calculations are best tabulated:

Frequency (Hz)	63	125	250	500	1000	2000	4000	8000
(i) L_2 from given data	70	79	81	83	80	81	79	75
(ii) L_1 from NR 35 (see *Fig 3*)	63	52	45	39	35	32	30	28
(iii) $10 \log_{10} S$ (where $S = 36$)	15.6	15.6	15.6	15.6	15.6	15.6	15.6	15.6
(iv) $10 \log_{10} A$	14.0	15.4	18.5	21.4	22.2	22.2	21.9	21.8
Total: (i) − (ii) + (iii) − (iv)	8.6	27.2	33.1	38.2	38.4	42.4	42.7	40.8

The total values given in the table show the minimum sound reduction index of the partition at each frequency so that the sound level in the office does not exceed the NR 35 curve.

Rating of industrial noise

The method of rating industrial noise affecting mixed residential and industrial areas is given in BS 4142. The method is intended for rating industrial noise affecting people living in the vicinity. It may also prove useful in planning new industrial buildings and alterations to existing buildings. In applying the method of BS 4142 the noise level, in dBA, produced by the industrial process is measured near the dwelling under consideration. This measured level is then corrected for various characteristics of the noise to produce the **corrected noise level**.

The background noise level, without the industrial noise, is then measured or if this is impossible a notional background noise level is calculated. If the corrected noise level exceeds the background noise level by more than 10 dBA then complaints may be expected. If it exceeds it by 5 dBA the situation is marginal. If the corrected noise level is more than 10 dBA below the background noise level then no complaints are likely to arise.

To determine the corrected noise level:

(a) If the noise is steady record the level, L_s, in dBA. If periods of louder noise or intermittent bursts occur measure this higher level, L_h, in dBA.

(b) Apply the following corrections:

(i) *Tonal character:* if noise has a definite distinguishable continuous note (whine, hiss etc.) add 5 dBA to measured value.

(ii) *Impulsive character:* if there are impulsive irregularities (bangs, thumps etc.) add 5 dBA to measured level.

(iii) *Intermittency:* if the noise is not continuous find the on-time duration and the sum of the on-times as a percentage of the total time. *Fig 4* shows the corrections for use during day and evening operation. BS 4142 also contains similar corrections for night-time operation.

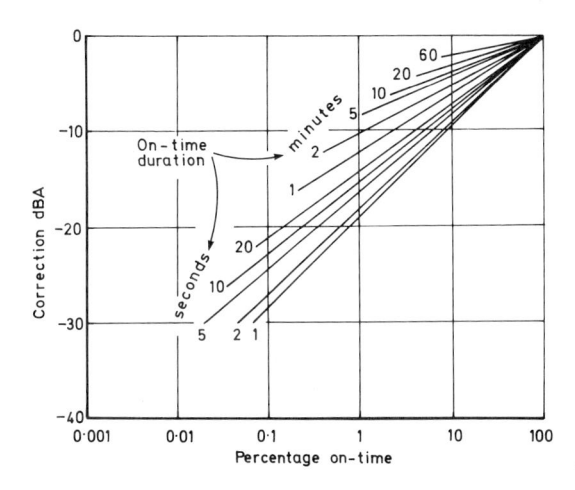

Fig 4 Intermittency and duration correction for day and evening

The corrections are applied separately to L_s and L_h to give the corrected noise levels L_s' and L_h'.

To determine the background noise level: this is measured, in dBA, in the absence of the industrial noise. If however this is impossible the notional background noise level is obtained by taking a value of 50 dBA and applying corrections for the type of installation, the type of district and the time of day. The reader is referred to BS 4142 for values of these corrections.

Problem 6 In the modification of a manufacturing process a company proposes to install a new machine with a dust extracting unit located externally to the factory. This unit draws in dusty air, extracts the dust, and blows the clean air to the outside which creates a whining noise. The unit will be operated only during the working day for a period of approximately 10 minutes in each hour. By consulting manufacturer's data and by studying a similar installation it is estimated that the sound level near adjacent houses will be 68 dBA. The existing background noise level near the houses is measured as 60 dBA. Determine whether complaints are likely.

Peak sound level L_h	= 68 dBA
Correction for tonal character	= +5 dBA
Correction for impulsive character	= 0
Correction for intermittency	= the on-time is 10 minutes in each hour which gives a percentage on-time of 16.7%. From *Fig 4* it will be seen that the correction is −3 dBA.
Corrected noise level = 68 + 5 − 3	= **70 dBA**.

This value exceeds the background noise level by 10 dBA and complaints are likely. Consideration should thus be given to reducing the noise from this unit before installation.

Noise from construction and demolition sites

There are two main aspects to be considered. Firstly the exposure to noise of persons on the site. This has been covered earlier in this chapter under the Code of Practice for Reducing the Exposure of Employed Persons to Noise. The second aspect is the effect of noise on the neighbourhood which is of concern to persons living and working near the site.

The Control of Pollution Act 1974 empowers local authorities to impose noise requirements if they so wish. The Act leaves local authorities to establish realistic noise limits in light of local needs. The limits could be set in terms of an equivalent continuous sound level, L_{eq}, based on A-weighted sound energy, a maximum sound level in dBA or hours during which the work may be carried out. The Act established a consent procedure whereby the contractor can approach the local authority to ascertain its noise requirements at the planning stage. It is not possible to discuss this in detail in a book of this nature but the reader may gain valuable advice from BS 5228: *Code of Practice for noise control on construction and demolition sites.*

Two acoustic aspects need consideration; these are the prediction of sounds pressure levels due to construction plant and the evaluation of L_{eq}.

The prediction of sound pressure levels, in the simplest case, assumes that the plant is standing on the ground which is smooth, and reflective. It is also necessary to know the sound power level of the machine. In all cases the values are to be A-weighted. Under these assumptions it may be shown that the sound pressure level at a distance from a machine is given by

$$L_{p(A)} = L_{w(A)} - 20 \log_{10} r - 8$$

where $L_{p(A)} = $ is the sound level in dBA at a distance r from the source and $L_{w(A)}$ is the A weighted sound power level of the source.

Problem 7 A crane has an A-weighted sound power level of 112 dBA. Determine the sound level in dBA at position which is a distance 25 m from the crane.

From above:
$L_{p(A)} = L_{w(A)} - 20 \log_{10} r - 8$
$= 112 - 20 \log_{10} 25 - 8$
$= 112 - 28 - 8 =$ **76 dBA**

In the above calculation it is further assumed that the sound source is uniform and that the point at which the sound level is found is not too near the facade of a building which would cause an increase in sound level due to reflections.

Problem 8 For the position considered in *Problem 7* determine the sound level if in addition to the crane there is a compressor, having an A-weighted sound power level of 108 dBA, at a distance of 12 m.

The sound level due to the compressor is found by:
$L_{p(A)} = L_{w(A)} - 20 \log_{10} r - 8$
$= 108 - 20 \log_{10} 12 - 8 = 78$ dBA

From *Problem 7* it is already known that the sound level due to the crane is 76 dBA. It is now necessary to add these two levels by the method given in chapter 5. For the 76 dBA level:

$76 = 10 \log_{10} \dfrac{p_A^2}{p_0^2}$ where p_A is the A-weighted sound pressure.

$\dfrac{p_A^2}{p_0^2} = 3.98 \times 10^7$

For the 78 dBA level in a similar manner:

$$\frac{p_A^2}{p_o^2} = 6.31 \times 10^7$$

The total value is:

$$\frac{p_A^2}{p_o^2} = 3.98 \times 10^7 + 6.31 \times 10^7 = 1.029 \times 10^8$$

The total sound level is given by

$$10 \log_{10} 1.029 \times 10^8 = \mathbf{80\ dBA}.$$

Alternatively the chart given in chapter 5 for the addition of decibels can be used.

The evaluation of L_{eq} is similar to the methods given earlier in this chapter but BS 5228 gives an alternative form for the equation which is:

$$L_{eq} = 10 \log_{10} \left(\frac{1}{T}(t_1 . 10^{L_1/10} + t_2 . 10^{L_2/10} + t_3 . 10^{L_3/10} + \ldots)\right)$$

where T is the period over which the value of L_{eq} is to be found and $L_1, L_2, L_3, \ldots$ are the sound levels in dBA maintained for periods of $t_1, t_2, t_3, \ldots$ hours.

This is only applicable if the durations of separate levels can be readily distinguished. If the sound is fluctuating, irregular, intermittent or impulsive then an integrating sound level meter giving a direct reading of L_{eq} will be necessary.

> *Problem 9* A concrete breaker is to be used at two positions on a demolition site. The A-weighted sound power level of this breaker is 116 dBA. The breaker will be used for 8 hours at a distance of 15 m from a nearby house followed by a 4 hour period at a distance of 30 m from the house. Determine the value of L_{eq} for this 12-hour period.

In the first position the sound level at the house is found using:

$$L_{p(A)} = L_{w(A)} - 20 \log_{10} r - 8$$
$$= 116 - 20 \log_{10} 15 - 8 = 84\ \text{dBA}$$

Similarly for the second position:

$$L_{p(A)} = 116 - 20 \log 30 - 8 = 78\ \text{dBA}$$

For finding the value of L_{eq} the formula given above is used in which T is the total duration of 12 hours. Hence:

$$L_{eq} = 10 \log_{10} \left(\frac{1}{12}(8.10^{84/10} + 4.10^{78/10})\right)$$

$$= 10 \log_{10} \left(\frac{1}{12}(8.10^{8.4} + 4.10^{7.8})\right)$$

$$= 10 \log_{10} \left(\frac{1}{12}(8 \times 2.51 \times 10^8 + 4 \times 6.31 \times 10^7)\right)$$

$$= 10 \log_{10} \left(\frac{1}{12} \times 2.26 \times 10^9\right)$$

$$= 10 \log_{10} 1.88 \times 10^8 = \mathbf{82.7\ dBA}.$$

Traffic noise and dwellings

Traffic noise fluctuates rapidly with time and much research has been carried out to find a suitable measure of the effect of traffic noise. The requirements of this measure are twofold. Firstly that it should correlate with human response to the noise and also

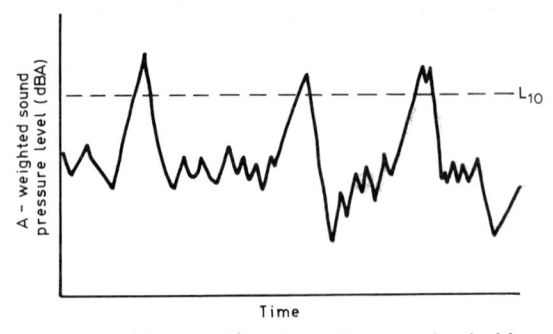

Fig 5 Variation of A-weighted sound pressure level with time, indicating value of L_{10}

be predictable in advance of the construction of new roads and motorways. The unit in present use is L_{10} which is defined as the sound level in dBA which is exceeded for one-tenth of the time. *Fig 5* shows the variation of A-weighted sound pressure level with time and the value of L_{10}.

The Noise Insulation Regulations cover traffic noise due to new road construction and significant changes in existing roads as it effects dwellings. The following is a very brief summary of these Regulations which the reader is advised to consult. The value of L_{10} in these regulations is the sound level in dBA, at 1 m from the facade of a dwelling, exceeded for one-tenth of a period of one hour between 6.00 and 24.00 hours on a normal working day. The value of L_{10} (18-hour) is the arithmetic average of all the levels of L_{10} during the period 6.00 and 24.00 hours on a normal working day.

The Regulations provide for compensation, in the form of insulation, where:
(i) that within 15 years of opening a new or altered highway the value of L_{10} (18 hour) will reach 68 dBA.
(ii) the traffic noise within the 15-year period will be at least 1.0 dBA higher than before work on the new or altered highway began.
(iii) that when the noise from the new or altered highway is added to the noise from any other highways in the vicinity, the total traffic noise is increased by at least 1.0 dBA.

The insulation that can be provided includes: double windows with a venetian blind for solar control if necessary, ventilator units, permanent suitably designed vents to provide air for combustion appliances and double doors.

The prediction of L_{10} (18-hour) in advance of construction is beyond the scope of this book and the reader is advised to consult Building Research Establishment Digests 185 and 186.

Problem 10 A dwelling stands near an existing road and the prevailing noise level is predicted on the basis of the value of L_{10} (18-hour) to be 70 dBA. A new road is to be built further away from the dwelling and parallel to the existing road which will in future carry local traffic. It is calculated that the maximum values of L_{10} (18-hour) within 15 years will be 66 dBA and 65 dBA due to the existing and new roads respectively. Determine whether the owner of the dwelling is entitled to insulation.

It is first necessary to calculate the relevant noise level after the construction has been completed. This is found by adding the sound levels of 66 dBA and 65 dBA due to the existing and new roads.

For the 66 dBa

$$66 = 10 \log_{10} \frac{p_A^2}{p_0^2}$$

whence $\frac{p_A^2}{p_0^2} = 3.981 \times 10^6$

Using the same method for the 65 dBA value

$$\frac{p_A^2}{p_0^2} = 3.162 \times 10^6$$

Thus the total value of $\frac{p_A^2}{p_0^2} = 7.143 \times 10^6$

Total sound level = $10 \log_{10} 7.143 \times 10^6$ = **68.5 dBA**

It will be noted that L_{10} (18-hour) after the new road is built will exceed 68 dBA but it will not be at least 1 dBA greater than the prevailing noise level; in fact it will be less since the traffic has been moved away from the dwelling. There will thus be no entitlement to insulation.

Perceived noise level and aircraft noise

It was seen, in chapter 5, that the loudness of a pure tone varies with frequency. The noisiness of a noise also depends on its frequency spectrum and is expressed by the

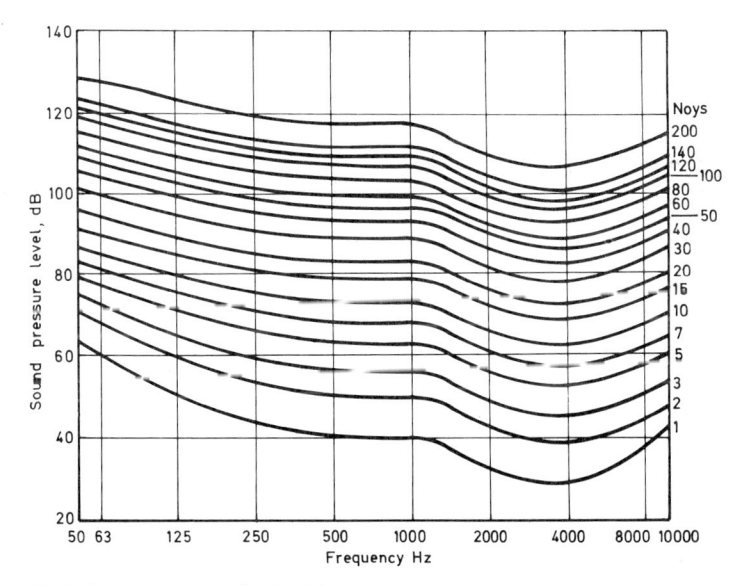

Fig 6 Contours of perceived noisiness

perceived noise level, L_{PN}, in dB. In many works the unit is denoted as PN dB. The evaluation of the perceived noise level is carried out in the following stages:
(i) Measure the sound pressure level in each of the one-third octave bands from 50 Hz to 10 000 Hz.
(ii) From *Fig 6* find the perceived noisiness, n, in noys for each of the sound pressure levels in (i).
(iii) Find the total noisiness, N, by using
$$N = n_{max} + 0.15\,(\Sigma n - n_{max})$$
where n_{max} is the greatest value of n and Σn is the total of all the values.
(iv) Calculate the perceived noise leve, L_{pN}, from:
$$L_{pN} = 40 + \frac{10 \log_{10} N}{\log_{10} 2}$$

Problem 11 Calculate the perceived noise level for a noise which has the following analysis:

Frequency (Hz)	50	63	80	100	125	160	200	250
Sound pressure level (dB)	63	64	64	66	67	70	80	85
Frequency (Hz)	315	400	500	630	800	1000	1250	1600
Sound pressure level (dB)	83	82	82	82	81	81	81	82
Frequency (Hz)	2000	2500	3150	4000	5000	6300	8000	10000
Sound pressure level (dB)	82	83	83	85	84	81	79	77

For each frequency use the value of the frequency and the sound pressure level to determine the value of n from *Fig 6* to give the following values:

Frequency	50	63	80	100	125	160	200	250
n	0.9	1.5	2.0	3.0	3.5	5.5	13.0	20.0
Frequency	315	400	500	630	800	1000	1250	1600
n	18.0	18.5	18.5	18.5	17.0	17.0	20.0	27.5
Frequency	2000	2500	3150	4000	5000	6300	8000	10000
n	31.5	39.0	41.5	48.0	41.5	31.5	22.5	16.0

It will be found that $\Sigma n = 475.9$ and $n_{max} = 48$.
The total noisiness using the formula in (iii) above is:
$$N = 48 + 0.15\,(475.9 - 48) = 112.2$$
The perceived noise level using the formula in (iv) above is:
$$L_{PN} = 40 + \frac{10 \log_{10} 112}{\log_{10} 2} = 108 \text{ dB}$$

The above method applies only when the broad band noise spectrum does not have pronounced irregularities. Where pronounced irregularities exist, such as a pure tone, then BS 5727 gives a method for tonal correction. The corrected value is called the tone-corrected perceived noise level and is denoted by L_{TPN}. Furthermore the total subjective effect of an aircraft flyover depends not only on the maximum value of L_{TPN} but also the duration of the flyover. BS 5727 gives a method for finding a duration allowance and the final corrected value is called the effective perceived noise level, denoted by L_{EPN}. In some works this is written as EPN dB.

The values of perceived noise level discussed above relate to a single flyover; the subjective effect of many flyovers depends also on the number of aircraft flyovers in a given time. Several methods exist for taking into account the number of aircraft. It is possible to calculate a continuous equivalent sound level, L_{eq}, using the perceived noise level values throughout the period. Much work has been done using the Noise and Number Index (NNI) which can be approximately defined as follows:

NNI = average of the maximum perceived noise levels
+ 15 $\log_{10}$ (number of aircraft flyovers) − 80

A detailed explanation of the calculation of NNI is beyond the scope of this book.

EXERCISES (answers on page 186)

In the following exercises where options are given select the correct option or options.

1 The sound pressure variations are transmitted from the drum membrane to the oval window by:
   (a) the pinna;
   (b) the ossicles;
   (c) the hammer, anvil and stirrup;
   (d) the cochlea;
   (e) the basilar membrane.

2 The auditory nerve is connected to:
   (a) the pinna;
   (b) the eardrum;
   (c) the bony ledge in the cochlea;
   (d) the ossicles.

3 The effect of a temporary threshold shift due to exposure to intense sound is:
   (a) permanent;
   (b) most noticeable at low frequencies;
   (c) recoverable after a period of time;
   (d) most noticeable at a frequency of 4000 Hz.

4 A permanent threshold shift:
   (a) is additive to presbycusis;
   (b) is recoverable if noise exposure is reduced;
   (c) due to old age but not exposure to noise;
   (d) spreads down into lower frequencies if exposure to noise persists.

5 The Code of Practice on Reducing the Exposure of Employed Persons to Noise suggests a maximum value of L_{eq} (8 hour) as:
   (a) 80 dBA; (b) 85 dBA; (c) 90 dBA; (d) 95 dBA.

6 Calculate the value of L_{eq} (8 hour) for the following noise exposure during a working day:
   95 dBA for 30 minutes;   87 dBA for 3 hours;
   91 dBA for 1 hour;   80 dBA for 3½ hours.

7 Find the maximum period of exposure to a sound of 105 dBA if L_{eq} (8 hour) is not to exceed 90 dBA, assuming that the sound level is below 85 dBA for the rest of the working day.

8   An office is separated from a production area by a partition of area 18 m². The sound pressure levels in the production area and the acoustic absorption of the office are given in the table below.

Frequency (Hz)	63	125	250	500	1000	2000	4000	8000
Sound pressure level (dB)	73	70	69	71	70	65	71	56
Acoustic absorption (m²)	7.2	8.1	10.8	13.0	13.7	13.0	13.7	16.6

(a) Determine the sound reduction index of the partition at each frequency if the sound level in the office is not to exceed NR 35.

(b) The mass per unit area of the partition (remembering that $R_{av} = 10 + 14.5 \log_{10} m$).

9   A bedroom has a volume of 63 m³ and the external wall, which has an area of 12.5 m² faces a main road. The sound pressure level outside the external wall is given in the table below:

Frequency (Hz)	63	125	250	500	1000	2000	4000	8000
Sound pressure level (dB)	81	77	73	69	65	64	59	45

Assuming that the reverberation time of the bedroom may be taken as 0.5 s determine the sound reduction index of the external wall at each frequency if the sound level in the bedroom is not to exceed NR 25.

10 Complaints are likely if the corrected noise level due to industrial noise near dwellings exceeds the background noise level by:
(a) 5 dBA;   (b) 10 dBA;   (c) 15 dBA;   (d) 20 dBA

11 In assessing the noise due to a new factory it was found that the maximum sound level was 65 dBA with a definite whine but no bangs or thumps. The noise was intermittent and occurred five times an hour with an on-time of 3 minutes each time. Find the corrected noise level.

12 A machine on a construction site has an A-weighted sound power level of 116 dBA. Determine the sound pressure level in dBA at a distance of 20 m from the machine.

13 At what distance from the machine specified in *Exercise 12* will the sound pressure level by 90 dBA.

14 Determine the sound pressure level in dBA at a point which is 18 m from a machine having an *A*-weighted sound power level of 110 dBA and 10 m from another machine having an *A*-weighted sound power level of 106 dBA. Assume that both machines are in use at the same time.

15 Find L_{eq} (12-hour) for the noise due to construction work which is 86 dBA for 5 hours followed by 80 dBA for 3 hours. For the rest of the 12 hour period the noise level is 65 dBA.

16 A dwelling stands near the junction of two roads. The prevailing noise levels are predicted on the basis of the value of L_{10} (18-hour) to be:
road A   64, dBA;   road B   65 dBA
Road B is to be significantly altered and the values of L_{10} (18-hour) within the next 15 years will become:
road A   64 dBA;   road B   67 dBA
Determine whether the owner is entitled to compensation.

17 The perceived noise level may be calculated from an octave band analysis by replacing the formula previously given for one-third octave analysis by:
$N = n_{max} + 0.3\,(\Sigma n - n_{max})$.

Use this to determine the perceived noise level for a noise with the following analysis:

Frequency (Hz)	63	125	250	500	1000	2000	4000	8000
Sound pressure level (dB)	69	73	89	87	86	87	91	82

9 Noise control

There are three possibilities in noise control; these are to modify the source of noise, control the sound in its path from source to receiver and control the sound at the receiver. It is not possible in this book to give a full treatment of noise control but to introduce the reader to some examples in construction.

NOISE CONTROL AT SOURCE

The following examples should illustrate the possibilities of modifying existing equipment or selecting a quieter method.

(a) *Dumpers.* The Building Research Establishment have shown that a worthwhile reduction in noise can be achieved by fitting efficient exhaust and inlet silencers and by enclosing the engine in a suitably designed enclosure to avoid engine overheating.

(b) *Concrete breakers.* For pneumatic hand held concrete breakers the main sources of noise are the compressor, the exhaust and the 'ring' of the tool on the concrete whilst it is in operation. The noise from the compressor can be reduced by a suitably designed enclosure and the use of a muffler will reduce the exhaust noise. Sound deadened tools are available where the 'ring' of the tool is reduced by a synthetic rubber ring which damps the vibration of the tool. The use of an electric or hydraulic breaker overcomes the problems of the air exhaust noise but attention must be paid to the noise produced by the electric generator, if used, and the hydraulic pumping system.

For larger work the Building Research Establishment designed the 'Nibbler' which attaches to a hydraulic excavator and snaps concrete in tension by applying a bending moment to it. The machine is suitable for breaking concrete roadways, runways, factory bases and similar applications.

(c) *Piling.* In some cases it may be possible to select a different method of piling for instance a jacked system for sheet piles rather than a hammer system.

(d) *Woodworking machines.* Circular saw blades produce considerable noise where idling due to resonant vibrations of the blade. This noise can be reduced by spring loaded dampers or by using integrally damped blades which also reduce the noise during cutting.

Cutterblocks with helical blades will produce less noise than with straight blades since the cutter is continuously in contact with the timber thereby reducing the impacts between the blades and the timber which occur with straight blades.

NOISE CONTROL IN TRANSMISSION OUTDOORS

(a) Distance

The careful positioning of noise sources away from noise sensitive areas can be of considerable value in reducing noise. When considering noise on construction sites in chapter 8 it was seen that for a point source on the ground the sound pressure level, the sound power level and the distance are related by:

$L_p = L_w - 20 \log_{10} r - 8$

This relationship allows an investigation of the change in sound pressure level with distance. The above equation may be used with A-weighted values as in chapter 8.

Problem 1 A machine has a sound power level of 100 dB. Evaluate the sound pressure level at distances of 5, 10, 20, 40 m from the machine.

In each case the sound pressure level is given by:

$L_p = L_w - 20 \log_{10} r - 8$

When the distance is 5 m then

$L_p = 100 - 20 \log_{10} 5 - 8 = 78$ dB

If the reader repeats this calculation for the distances of 10, 20 and 40 m the following sound pressure levels will be found: **72 dB, 66 dB and 60 dB**. It will be seen that the sound pressure level decreases 6 dB each time the distance is doubled.

The absorption of sound by the air molecules may be of importance where the distances involved are very large or the sound is of high frequency. Wind and temperature gradients will affect the sound level but the magnitude of these effects is not readily calculable.

(b) Machinery enclosures

These should possess adequate sound insulation so that the sound does not pass through them and be lined on the inside with an absorbent material so that the sound level inside does not build up thereby effectively increasing the acoustic power of the source.

The enclosure should be as complete as possible but a certain amount of openings is usually necessary for pipes, wires, ventilation and access. Although a maximum reduction of 50 dB is possible this is not usually necessary and a carefully constructed enclosure using a material of surface mass 10 kg/m² or more should be adequate.

(c) Acoustic sheds

Noisy processes may be performed in an acoustic shed which consists of an enclosure with one side open and preferably a moveable screen in front of the open side.

For best results the inside surfaces of both the shed and the screen should be lined with an absorbent material. *Fig 1* shows some typical reduction figures for sheds lined with absorbent material.

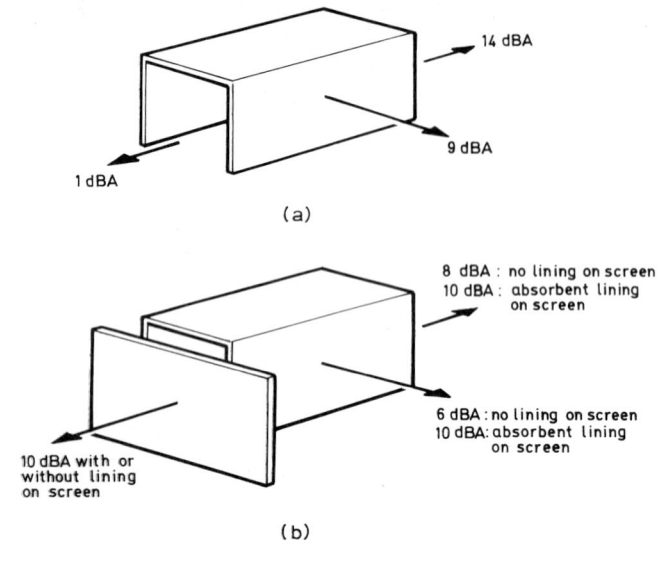

Fig 1 Sound reduction by absorbent lined acoustic sheds (a) with no screen; (b) with screen with and without absorbent lining

(d) Barriers

When a barrier is placed in the path of the sound a shadow zone is created as shown in *Fig 2*. Unlike light the shadow is not at all sharp edged. Diffraction of sound at the top edge of the barrier causes sound to enter the shadow zone. The attenuation of the sound depends on the Fresnel number $N = 2d/\lambda$ where d is the path difference illustrated in *Fig 2* and λ is the wavelength of the sound.

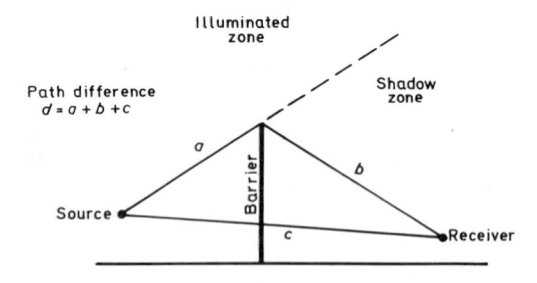

Fig 2 Path difference due to an acoustic barrier

Fig 3 shows the likely sound attenuation for a long barrier for values of the Fresnel number up to about 10; for values above this reflections, which always occur in practice, will limit the attenuation to a value of about 24 dB.

Problem 2 A noise source 1 m above the ground produces a sound pressure level of 75 dB at a frequency of 500 Hz at a receiver 16 m away at a height of 2 m above the ground. A long barrier, of height 3 m, is placed between the source and the receiver.

Find the sound pressure level at the receiver if the barrier is (a) 2m from the source (b) 8m from the source. The velocity of sound may be taken as 340 m/s.

In the first instance it is necessary to find the wavelength of the sound, λ. In chapter 5 it was found that $v = f\lambda$ which on rearrangement gives:

$$\lambda = \frac{v}{f} = \frac{340}{500} = 0.68 \text{ m}$$

To find the path difference, d, the theorem of Pythagoras is applied to the situation shown in *Fig 3*:

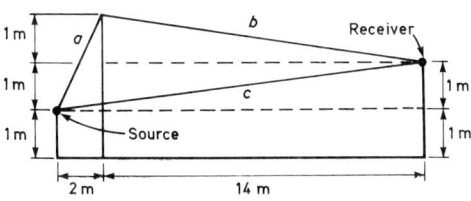

$a = \sqrt{2^2 + 2^2} = \sqrt{8} = 2.828$
$b = \sqrt{1^2 + 14^2} = \sqrt{197} = 14.036$
$c = \sqrt{1^2 + 16^2} = \sqrt{257} = 16.031$

Fig 3 Diagram for Problem 2(a)

The path difference d is then found by:

$d = a + b - c = 2.828 + 14.036 - 16.031 = 0.833$

The Fresnel number is:

$$N = \frac{2d}{\lambda} = \frac{2 \times 0.833}{0.68} = 2.45$$

From *Fig 4* it is found that the attenuation is 17 dB so that the resulting sound pressure level at the receiver is:

sound pressure level = 75 - 17 = **58 dB**

The answer for part (b) is obtained in a similar manner; the reader is requested to check the following calculations

$a = \sqrt{8^2 + 2^2} = \sqrt{68} = 8.246$
$b = \sqrt{1^2 + 8^2} = \sqrt{65} = 8.062$
$c = \sqrt{1^2 + 16^2} = \sqrt{257} = 16.031$
$d = 8.246 + 8.062 - 16.031 = 0.277$

The Fresnel number is:

$$N = \frac{2d}{\lambda} = \frac{2 \times 0.277}{0.68} = 0.81$$

From *Fig 3* the attenuation due to the barrier is about 12 dB so that the resulting sound pressure level = 75 - 12 = **63 dB**.

It will be seen that the barrier is most effective when it is placed near the source or the receiver. When sound reflected from the barrier towards the source

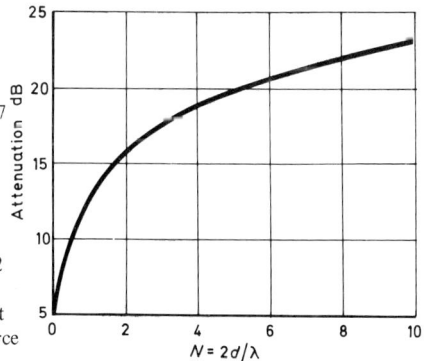

Fig 4 Attenuation of sound pressure level due to a barrier

causes an unacceptable increase in the sound pressure level near the source then the source side of the barrier must be treated with an absorbent material. If the barrier is not long then the sound pressure level due sound diffracted round the edges must be calculated and the resulting levels added.

NOISE CONTROL INDOORS

The sound field indoors consists of the direct field and the reverberant sound field as already discussed in chapter 6. The method of control adopted will depend on the relative magnitudes of the two components.

(a) Control of direct sound level

The methods applicable are the same as those for outdoors; these are to increase the distance between the source and receiver and to use barriers or screens.

The scope inside a room for increasing the distance is generally rather limited. Barriers and screens can be effective but care must be taken to line the source side of the screen with absorbent material so that reflected sound does not increase the sound pressure level on the source side.

If these methods are not successful then an enclosure will be necessary for the noise source.

(b) Control of reverberant sound field

In chapter 6 it was seen that the sound intensity in the reverberant field is given by:

$$I = \frac{4W}{R_c} \text{ where } R_c = \frac{S\bar{\alpha}}{(1-\bar{\alpha})}$$

It will be evident that if the average absorption coefficient, $\bar{\alpha}$, is increased then R_c will increase and the intensity, I, will decrease leading to a reduction in the sound pressure level. The reduction in sound pressure level due to increasing the absorption can be approximately estimated by the use of *Fig 5* where $\bar{\alpha}_1$, and $\bar{\alpha}_2$ are the average absorption coefficients before and after room treatment.

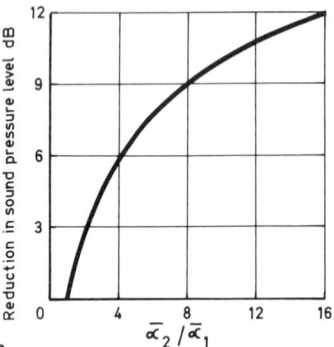

Fig 5 Reduction in sound pressure level due to increased absorption

Problem 3 A small workshop is 15 m by 10 m by 4 m high. The reverberant sound pressure level is measured as 80 dB. The surfaces of the room are all hard materials and it is decided to attempt to reduce the sound pressure level by treating the ceiling with acoustic tiles having an absorption coefficient of 0.55.

The reverberation time of the workshop is found to be 3.5 s and reference to tables indicates that the absorption coefficient of the existing ceiling will be 0.02.

Estimate the sound pressure level after treatment.

Before proceeding note that:
the volume of the room = 600 m^3
the surface area of the room = 500 m^2
the area of the ceiling = 150 m^2
The average absorption coefficient before treatment can be found using Sabine's formula:

$$T = \frac{0.16\ V}{A}$$

By rearrangement:

$$A = \frac{0.16\ V}{T} = \frac{0.16 \times 600}{3.5} = 27.43$$

The average absorption coefficient, $\bar{\alpha}_1$, is given by:

$\bar{\alpha}_1 = \frac{A}{S}$ where S is the area of the room surfaces

$$\bar{\alpha}_1 = \frac{27.43}{500} = 0.055.$$

When the ceiling is tiled the absorption in the room will be:

absorption = existing absorption + new absorption of ceiling − old absorption of ceiling covered by tiles.

$$= 27.43 + 150 \times 0.55 - 150 \times 0.02$$
$$= 106.93$$

Thus the average absorption coefficient, $\bar{\alpha}_2$, after treatment is:

$$\bar{\alpha}_2 = \frac{A}{S} = \frac{106.93}{500} = 0.21$$

To use *Fig 5*:

$$\frac{\bar{\alpha}_2}{\bar{\alpha}_1} = \frac{0.21}{0.055} = 3.8$$

It will be seen from *Fig 5* that the reduction in sound pressure level will be nearly 6 dB. Thus the sound pressure level after treatment will be 80 − 6 = **74 dB**.

NOISE CONTROL AT THE RECEIVER

When all other attempts at noise control are inadequate, treatment at the receiver must be used. For a human receiver there are two possibilities; firstly to reduce the period of exposure and secondly to provide ear plugs or ear defenders. If ear defenders are employed it is essential that they are chosen to provide the correct degree of protection and that they are maintained in good condition and, furthermore, that they are correctly used.

EXERCISES (answers on page 186)

In the following exercises where options are given select the correct option or options.

1   A machine is 15 m from a dwelling and creates a sound pressure level near the dwelling of 70 dBA. If the machine is moved to a distance of 30 m from the dwelling the sound pressure level would become:
    (a) 35 dBA; (b) 61 dBA; (c) 64 dBA; (d) 67 dBA.

2 If a machine is enclosed in an open sided acoustic shed with a screen, both of which are lined with absorbent material, the maximum reduction in sound pressure level for a receiver some distance away is likely to be:
   (a) 3 dBA; (b) 6 dBA; (c) 10 dBA; (d) 20 dBA.

3 A compressor having an A-weighted sound power level of 109 dBA is situated 10 m from a noise sensitive position. In order to reduce the noise two possibilities are to be investigated. Firstly it is possible to move the machine so that it is 25 m from the receiving position. Secondly an absorbent lined open sided shed and screen are available. Find the sound pressure level at the receiving point in the following cases:
   (i)   machine in original position;
   (ii) machine in original position but placed in acoustic shed;
   (iii) machine is moved to new position;
   (iv) machine is moved to new position and placed in acoustic shed.

4 An extractor fan is at a height of 1.0 m above ground and creates a sound pressure level of 65 dB at the first floor windows of a house which is 30 m away from the fan. The windows are 4 m above ground level. The predominant frequency produced by the fan is 250 Hz. Estimate the sound pressure level at the house is a long brick wall is built between the fan and the house. The wall is to be 3 m high and 2 m from the fan. Assume velocity of sound is 340 m/s.

5 The reverberant sound pressure level in a room is 70 dB. If the average absorption coefficient of the room surfaces is doubled then the sound pressure level will become:
   (a) 35 dB, (b) 64 dB; (c) 67 dB; (d) 76 dB.

6 A small workshop is 7 m × 7 m × 3 m high and has hard surfaces of plaster and concrete with an average absorption coefficient of 0.05. This workshop was converted to an office for six typists and complaints were soon received that it was noisy and reverberant. In an attempt to improve the acoustics, the floor was carpeted and acoustic tiles fixed on the ceiling and top 1 m of the walls. Estimate
   (i)   the reverberation time before treatment;
   (ii) the reverberation time after treatment;
   (iii) the reduction in reverberant sound pressure level created by the treatment.
   The following absorption coefficients may be assumed:
   typist 0.4 m², carpet 0.3, acoustic tiles 0.55. In each calculation remember to include the typists.

10 Vision, lighting units, colour and light sources

THE EYE AND MECHANISM OF VISION

Fig 1 shows a sectional diagram of the human eye. The cornea is a transparent window through which light enters the eye. The lens focuses the light entering the eye upon the retina at the back of the eye. The amount of light entering the eye is controlled by the iris which is a coloured diaphragm. The hole in the iris, called the pupil, decreases in diameter as the brightness of the light increases. Light sensitive receptors in the retina send signals via the optic nerve to the brain giving the sensation of vision.

There are two types of receptors: **rods** and **cones**. Rods are not sensitive to colour and operate in dim light and give scotopic vision. Cones which are sensitive to colour operate in bright light and give photopic vision.

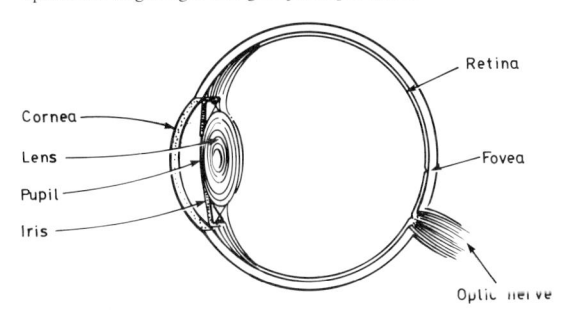

Fig 1 Simplified cross-section of human eye

The area of the retina called the fovea consists entirely of cones each one of which is individually linked to a fibre in the optic nerve. On moving away from the fovea, the density of cones decreases and the density of rods increases. In this region several receptors are connected to a single nerve fibre. Thus the fovea is used to produce clear detailed colour vision but the peripheral regions are very sensitive to any change and will often detect a flicker or movement which would otherwise pass unnoticed.

LIGHT

The portion of the electromagnetic spectrum with a wavelength from about 380 mm up to 700 nm gives rise to the sensation of light. The eye distinguishes different wavelengths as different colours. The short wavelengths give the violets and blues whereas

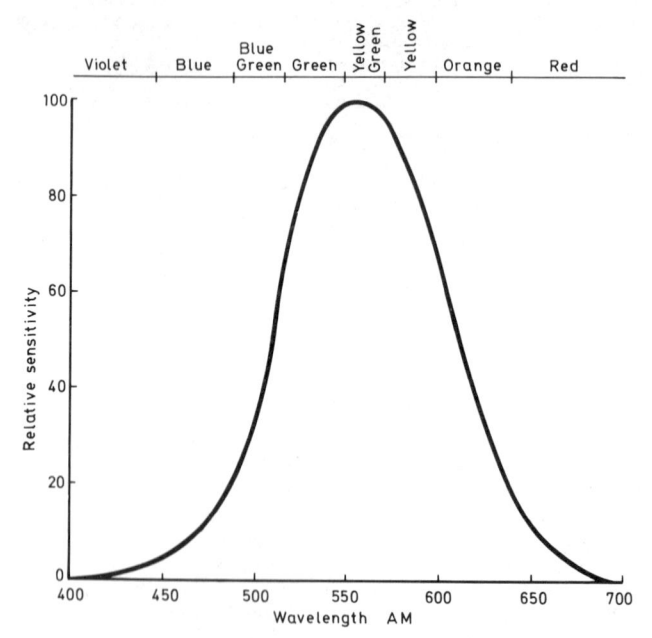

Fig 2 Variation of relative sensitivity of human eye to different wavelengths of light

the long wavelengths create the oranges and reds, these colours and wavelengths are shown in *Fig 2*.

The sensitivity of the human eye varies with the wavelength of the light. *Fig 2* shows the variation in sensitivity with wavelength for a light adapted eye, that is photopic vision. The peak sensitivity occurs at 555 nm and is a yellow green colour. The curve for scotopic vision is slightly different, having its peak shifted slightly to the left.

LIGHTING UNITS

Since the sensitivity of the human eye varies with wavelength the units of lighting have to take this into account. The following four quantities need to be considered:

(i) **Luminous flux** (symbol: F; Unit: **lumen (1 *l*m)**)
Luminous flux is the rate of flow of energy evaluated in accordance with the relative sensitivity of the eye. 1 watt of energy of wavelength 555 nm, which is the wavelength of the peak of the relative sensitivity curve, produces a luminous flux of 680 lumens. If the 1 watt of energy had a wavelength of 500 nm then the number of lumens is $0.3 \times 680 = 204$ lumens, since, as can be seen from *Fig 2*, the relative sensitivity of the eye at this wavelength is approximately 0.3.

(ii) **Intensity** (Symbol: I; Unit: **Candela (cd)**)
The amount of luminous flux from a light source does not give information about the

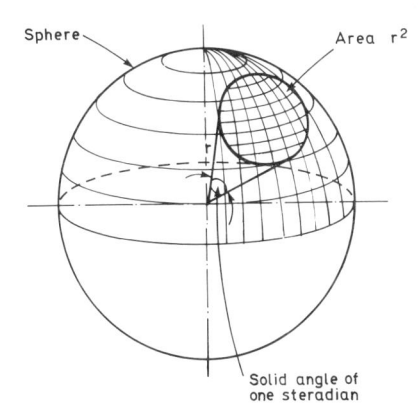

Fig 3 Unit solid angle

amount flowing in a particular direction. The rate of flow of luminous energy in a specified direction is termed the intensity and is stated in candela. In order to appreciate the relationship between luminous flux and intensity it is necessary to understand the concept of a solid angle. The unit of solid angle is the **Steradian** which is the solid angle at the centre of a sphere of radius r subtended by an area of r^2 on its surface, see *Fig 3*. Thus if a flux of F lumens is emitted into a solid angle ω the intensity I is given by:

$$I = \frac{F}{\omega}$$

If a point source which emits luminous flux equally in all directions is considered then the intensity in any direction is given by

$$I = \frac{F}{4\pi}$$

since there are 4π steradians in the complete sphere.

(iii) Illuminance (Symbol: E; Unit: Lux (lx))

When the luminous flux strikes a surface it is natural to consider the luminous flux per unit area, this is termed the illuminance. The units of illuminance are lumens per square metre to which the special name of lux is given. The illuminance, E, is related to the flux, F, and the area of the surface, A, by:

$$E = \frac{F}{A}$$

(iv) Luminance (Symbol: L, Unit: Candela per square metre (cd/m^2))

As objects are only visible by the amount of light which they emit or reflect in the direction of view this quantity is of considerable importance and is termed the luminance. The unit of luminance is the candela per square metre. For surfaces that emit light the luminance, L, is related to the intensity in a given direction, I, and the projected area in that direction, A, by:

$$L = \frac{I}{A}$$

For surfaces which reflect light the reflection characteristics of the surface are important. Diffusely reflecting surfaces reflect light uniformly in all directions. Such reflection is obtained from perfectly matt surfaces although many interior decorative finishes in building approximate to this. If the illuminance of the surface is E lux and the surface has a reflectance, R, then the amount of reflected light is RE. This light is reflected in all directions by a matt surface and it may be shown that the luminance for an observer at a particular point is given by:

$$L = \frac{RE}{\pi}$$

For surfaces which are not matt the reflectance depends on how the surface is illuminated and particularly on the direction of the incident light. In this case the luminance factor is used to quantify the reflection characteristics and refers to the total light reflected by specular and diffuse reflection.

In the practical implementation of the measurement of the above quantities the fundamental SI unit is the candela which is defined as the luminous intensity, in the perpendicular direction, of a surface of area 1/600 000 square metres of a black body at the temperature of the freezing point of platinum under a pressure of 101 325 Pa. It should be noted that the freezing point of platinum is approximately 1772°C. The definition of the lumen follows directly as the luminous flux emitted by a source of intensity 1 cd into a solid angle of 1 steradian.

Thus it will be seen that the units of lighting are defined in terms of a standard light source. This has been the case throughout the history of lighting units when the original source used was a standard candle. The relative sensitivity curve of the human eye is very useful in calculating the luminous flux where the energy spectrum is known and in designing equipment to measure illuminance and luminance.

Problem 1 A uniformly diffusing sphere of diameter 0.2 m emits a luminous flux of 1800 lumens. Determine (i) the intensity in any direction (ii) the luminance of the sphere.

For part (i) it will be remembered that for a source radiating light uniformly in all directions the intensity is given by:

$$I = \frac{F}{4\pi} = \frac{1800}{4\pi} = 143.2 \text{ cd}$$

For part (ii) the luminance is given by:

$$L = \frac{I}{A}, \text{ where } A \text{ is the projected area.}$$

In this case the projected area is a circle of area given by: $\pi(0.1)^2$. Thus the luminance is:

$$L = \frac{143.2}{\pi(0.1)^2} = 4560 \text{ cd/m}^2$$

Problem 2 The illuminance of a matt painted surface is 350 lux. If the reflectance of the paint is 0.7 find the luminance of the surface

The luminance of the surface is given by:

$$L = \frac{RE}{\pi} = \frac{0.7 \times 350}{\pi} = 78 \text{ cd/m}^2$$

It is to be noted that in many works the luminance of a surface is given as $L = RE$ and the unit used is the apostilb.

TYPES OF ELECTRIC LAMP

(i) Incandescent Lamps

An incandescent lamp consists of a coiled tungsten filament contained in a glass bulb which is filled with an inert gas. Since light is produced by heating the filament, much of the electrical energy supplied is dissipated as heat. Typically about 6% of the input energy is converted to light and the remainder is almost entirely radiated as infra red energy or is conducted and convected away from the lamp.

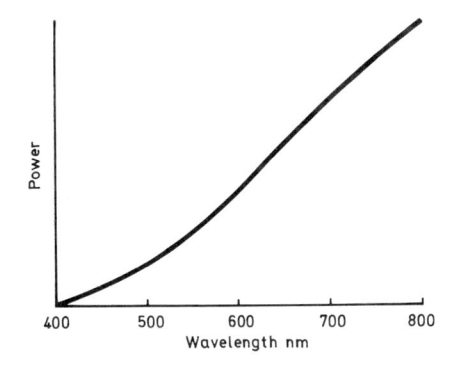

Fig 4 Spectral distribution of an incandescent lamp

The spectral distribution of the light emitted is biased to the red end of the spectrum as shown in *Fig 4*. Since the relative sensitivity of the human eye is low at the red end of the spectrum the luminous efficacy of the lamp is low. A luminous efficacy of 12 lumens/watt is typical, although this does depend on the type and wattage of the lamp. A lamp life of 1000 hours is typical for general lighting service (GLS) lamps.

(ii) Tungsten halogen lamps

A halogen, such as iodine is added to the gas filling used in a tungsten filament lamp. The halogen is used to reduce the rate of evaporation of the filament which occurs with the ordinary filament lamp. This permits the filament to be run at a higher temperature giving more light, longer life and reducing the blackening of the glass bulb due to the depositon of the evaporated tungsten.

The action of the halogen is as follows: near the lamp wall the halogen combines with the tungsten to produce a tungsten halide which is carried back to the filament by convection. At the high temperatures near the filament the tungsten halide disassociates releasing the halogen to repeat the cycle and returning the tungsten to the filament.

To prevent the tungsten halide being deposted on the walls of the bulb a minimum wall temperature of 250°C is essential. A small bulb is thus necessary, and to withstand the thermal shock the bulb is made of quartz. Dirt on the bulb, such as grease due to handling, leads to deterioration of the quartz thus reducing the useful life of the lamp.

Care is thus necessary in handling these lamps. Since the filament runs at a higher temperature than in an ordinary tungsten filament lamp, the light produced is whiter and a luminous efficacy in the range 15 to 35 lumens/watt is achieved. A luminous efficacy of about 20 luments/watt is typical.

(iii) Discharge lamps

In these lamps, light is produced by striking an arc between electrodes contained in a tube filled with a gas or metallic vapour. The best known of these lamps are the sodium and mercury lamps.

Since sodium and mercury do not vaporise at ordinary temperatures the discharge is started in an inert gas such as argon in the case of the mercury lamp. Once the discharge is started the current in the discharge will increase indefinitely unless it is limited by an electrical ballast. Thus all discharge lamps have a starter to initiate the discharge and a ballast to control it. To protect the discharge tube from damage and to maintain it at the correct temperature it is usual to enclose it in an outer glass bulb. The light emitted is at specific wavelengths and gives a line spectrum.

The spectrum of a low pressure sodium lamp consists almost entirely of yellow light at a wavelength of about 589 nm. This is near the peak of the relative sensitivity curve for the human eye giving a high luminous efficacy which may be as high as 160 lumens/watt.

Since the spectrum is monochromatic the colour rendering is very poor. By increasing the sodium vapour pressure the spectrum is broadened and the identification of colours becomes possible by the golden white light produced.

Mercury lamps have more lines in the spectrum; as can be seen in *Fig 5*. It should be noted that the principal lines occur in the ultra-violet, violet, green and yellow. The colour rendering particularly for red colours is rather poor.

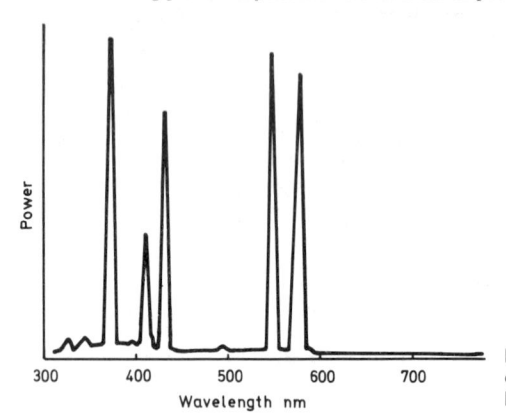

Fig 5 Spectral distribution of a mercury discharge lamp

Several methods are employed to improve the characteristisc of mercury lamps. Suitable phosphors applied to the inside of the outer glass bulb can be used to convert the ultra-violet radiation to visible light at a suitable wavelength thus extending the spectrum at the red end and improving the colour rendering. The combination of a mercury discharge and a tungsten filament, known as a blended lamp, along with the use of a phosphor on the inside of the outer glass bulb will improve the colour rendering still further.

The addition of metal halides to the mercury lamp together with the phosphor on the glass bulb will give very good colour rendering so that the lamp is suitable for use in large offices and shops as well as for industrial applications where good colour rendering is required.

(iv) Fluorescent lamps

At low mercury vapour pressures the mercury discharge produces a much higher proportion of ultra-violet radiation. In the fluorescent lamp, the inside of the glass envelope is coated with phosphor which converts the ultra-violet radiation to visible light. By using a suitable combination of phosphors virtually any desired colour can be obtained.

Luminous efficacies up to about 80 lumens per watt can be achieved but the value does depend on the tube size, wattage and colour. The spectrum of the light produced consists of a continuous spectrum having discrete peaks; a typical spectrum is shown in *Fig 6*.

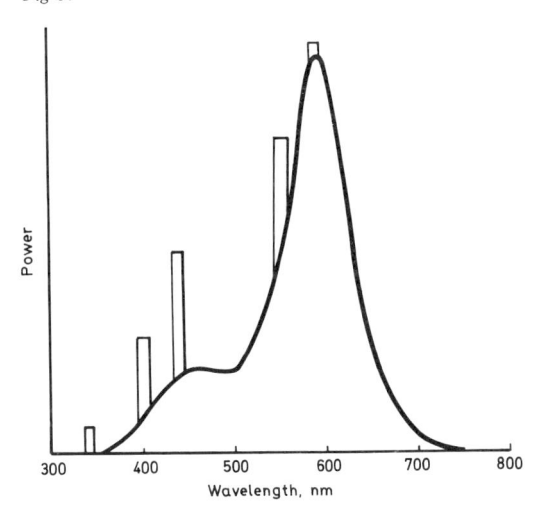

Fig 6 Spectral distribution of a 'white' fluorescent lamp

The ordinary fluorescent lamp is designed to operate with a surface temperature of $40°C$ in a corresponding ambient temperature of $25°C$. For temperatures above this, the light output decreases. Fluorescent tubes containing a ring of the metal indium which forms an amalgam with the mercury when the tube is in operation have their maximum efficiency at temperature above the normal values given above. These tubes can be used at higher temperatures which are likely to occur inside certain patterns of luminaire.

COLOUR OF LIGHT SOURCES

Several methods exist for describing the colour and appearance of light sources. Each method is outlined in the following paragraphs.

(i) Colour temperature

If a body is heated, as for example, the filament in a lamp, it will initially glow dull red. As the temperature is raised the light emitted will become whiter, since more light is emitted at shorter wavelengths. This effect has already been illustrated by *Fig 1* in Chapter 4.

For sources which have continous spectra, the colour can be related to that of a full radiator at a known temperature. For instance an ordinary filament lamp has a colour temperature of about 2700°K whereas natural daylight has a colour temperature around 6000°K.

For sources that do not have a continuous spectrum, a correlated colour temperature can be produced. For white fluorescent tubes the correlated colour temperature gives an indication of their colour appearance. As an example, the correlated colour temperature of a Warm White tube is about 3000°K and that of an Artificial Daylight tube is about 6500°K. For white sources the colour appearance becomes cooler as the colour temperatures increases. The following lamp colour appearances are often used:

WARM : correlated colour temperature up to 3000°K
INTERMEDIATE : correlated colour temperature between about 3000°K and 4500°K
COOL : correlated colour temperature usually about 6500°K.

The C.I.B.S Interior Lighting Code gives guidance on the choice of colour appearance of light sources for a wide range of applications.

(ii) Chromaticity coordinates

White light can be produced by the addition of three primary colours. Originally the primary colours used were red, green and blue. The mixture of two primaries produces a secondary colour. These mixtures can be represented by a colour triangle as shown

Fig 7 Red, green and blue colour triangle for additive colours

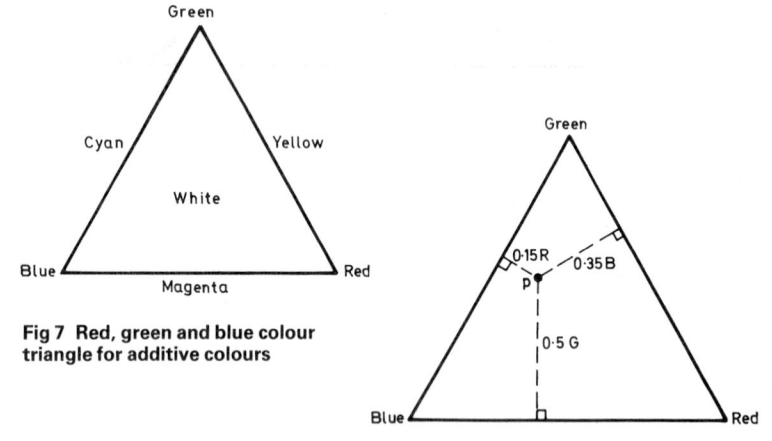

Fig 8 Specification of a colour within the colour triangle

in *Fig 7*. The secondary colours are magenta, cyan and yellow. Thus any colour can be specified by the perpendicular distances of the point from the opposite side of the triangle. A colour at point P in *Fig 8* is represented by:

Colour at P = 0.15 Red + 0.35 Blue + 0.5 Green

The colour system

It is an experimental fact that any spectral wavelength excluding the primaries cannot be matched by the addition of the primaries red, green and blue. The spectral wavelengths and quite a large number of colours lie outside the red, green, blue colour triangle. The colour at Q, in *Fig 9*, will require a negative amount of red. This appears to be unreasonable but it implies that the colour at Q plus an appropriate amount of red can be matched by a suitable combination of green and blue.

Fig 9 Specification of a colour outside the colour triangle

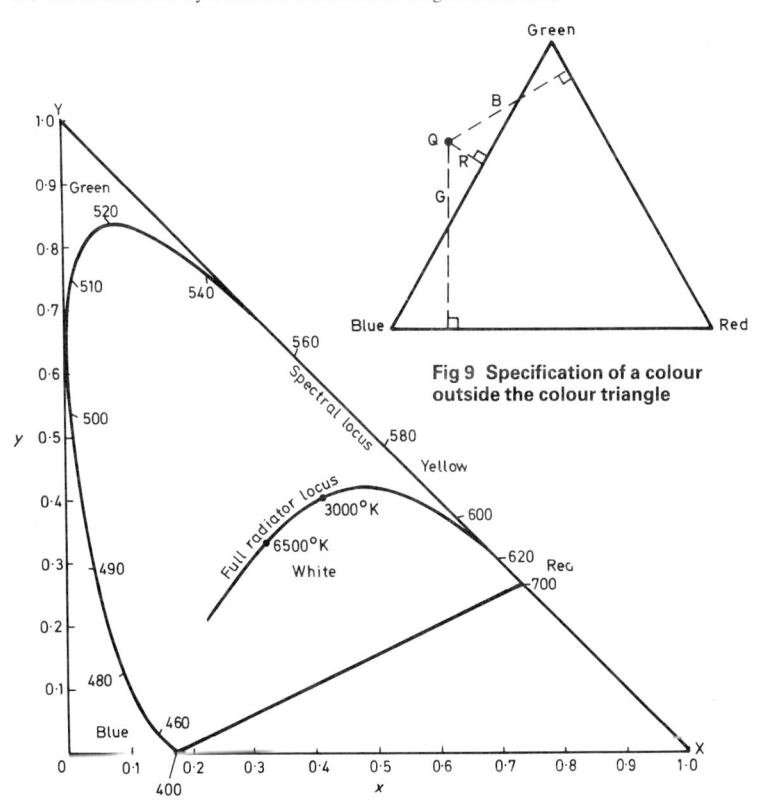

Fig 10 CIE Chromaticity diagram

To overcome the use of negative quantities in specifying colours the CIE (Commision Internationale de l'Eclairage) introduced three unreal primaries X, Y and Z. The colour triangle so produced now includes all the spectral colours which lie on the spectral locus shown in *Fig 10*. All other colours lie inside the spectral locus.

The colour system was devised so that for any colour the amounts x, y and z, of the primaries X, Y and Z required to match the colour satisfy the relationship $x + y + z = 1$. Thus it is only necessary to specify two chromaticity coordinates, the values of x and y usually being quoted. The locus of full radiators is also shown in *Fig 10*, some typical colour temperatures being indicated.

SURFACE COLOURS

The colour of a surface is produced by subtracting wavelengths from the incident light. Thus a surface that appears red under white light reflects light wavelengths at the red end of the spectrum and absorbs the other wavelengths. Surface colours are produced by a subtractive process.

Fig 11 shows the reflectance characteristics for a blue and a yellow pigment; the only region in which both pigments reflect is in the green. If the two pigments are mixed the resultant colour is green, this is to be compared with the additive process for light sources where blue and yellow produce white light. This occurs since yellow is obtained by the addition of red and green. The primary colours in the subtractive colour system are cyan, magenta and yellow. The colours obtained by mixing these primaries are shown in the colour triangle in *Fig 12* which should be carefully compared with the additive colour triangle in *Fig 7*.

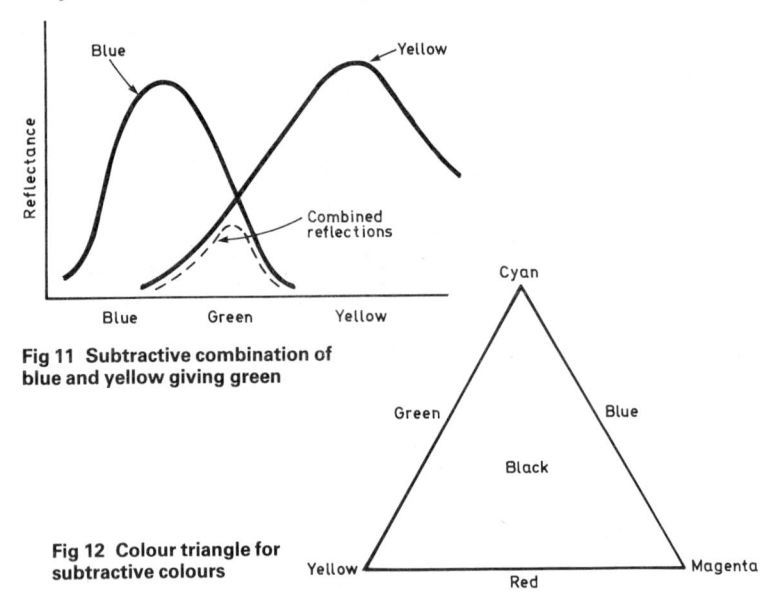

Fig 11 Subtractive combination of blue and yellow giving green

Fig 12 Colour triangle for subtractive colours

The specification of colour for surfaces is not easy and a number of methods have been adopted. Some of these will be briefly outlined.

(i) Munsell system classifies colour in three terms:

Hue: This describes the actual colour for example yellow or red. There are ten hues in the system: B, BG, G, Y, YR, R, RP, P, PB where B, G, Y, R and P stand for blue, green, yellow, red and purple respectively. The hue is preceded by a number, for example 5Y which is a mid yellow.

Value: This is a measure of the whiteness or greyness of the colour. This is perhaps best understood by imagining mixing a colour pigment with a succession of base colours ranging from white, through grey to black. There are ten divisions of value from pure white to pure black. 0 is pure black and 10 is pure white.

Chroma: This is a measure of the intensity of the colour, as an example it is easy to envisage the increase in chroma as more and more colour pigment is added to a white base. A chroma of 1 is a tint or shade which is almost neutral grey. A high chroma indicates a high richness of colour.

A typical Munsell specification might be 5Y 9/13 this being a mid-yellow with a value of 9 and a chroma of 13. This colour will be a bright, fairly intense mid yellow. A specification of 5Y 6/2 will be much greyer due to the lower value and considerably less colourful due to the low chroma. The reflectance of a surface is approximately related to its Munsell value as follows:

$R = V(V - 1)$

where R is the percentage reflectance of white light and V is the Munsell value. For instance for the specification 5Y 6/2 the value $V = 6$ and hence the reflectance $R = 30\%$.

(ii) BS 5252: *Framework for Colour Coordination for Building Purposes* gives a range of colours, the identification of each colour being in terms of hue, greyness and weight. The hues are specified by the following numbers:

00 neutrals
02 red-purposes
04 reds
06 yellow-reds
08 yellow-reds
10 yellows
12 green-yellows
14 greens
16 blue-greens
18 blues
20 purple-blues
22 violets
24 purples

The greyness is denoted by the letters A to E from maximum greyness to zero greyness.

Weight is a subjective form for lightness modified to produce colours of the same character in different hues. The weight is given by an odd number from 01 to 55. A typical specification might be 10E53. The '10' indicates a yellow, 'E' indicates zero greyness and '53' indicates the weight; the colour is a vivid yellow. British Standards derived from this standard give coordinated colour ranges for paints (BS 4800), vitreous enamesl (BS 4900), plastic (BS 4901), sheet and tile flooring (BS 4902).

(iii) A colour can be specified by its CIE chromatacity coordinates and its luminance factor. For the 10E53 colour mentioned above the chromaticity coordinates are $x = 0.4406$; $y = 0.4882$ and the luminance factor is 64.6 per cent.

COLOUR RENDERING

It is well known that the apparent colour of a surface may change under different sources of illumination. An extreme example is that of a surface which is blue in daylight will appear grey in sodium light.

There are a number of possible methods of attempting to quantify the colour rendering properties of light sources. The CIE colour rendering index refers to the accuracy with which colours are reproduced by a given light source relative to their apperance under a reference source. A lamp is given a general colour rendering index denoted by R_a, which may be supplemented by special colour rendering indices.

The CIE index has a number of limitations and has not so far been widely adopted in Great Britain. The CIBS Code for Interior Lighting gives valuable guidance on the type of lamps that are suitable in a wide range of applications. As an example the Code recommends for colour matching in paint works fluorescent lamps having a cool appearance and being Northlight or Artificial Daylight in colour.

Surface colour metamerism can occur when two different surfaces match in colour under one illuminant but no longer match under a different illuminant which has a different spectral distribution. This can arise where the materials have different dyes or pigments such as dyed cloth and coloured plastic.

EXERCISES (answers on page 186)

In the following exercises where options are given select the correct option or options.

1. In the human eye rods are:
   (a) sensitive to colour;
   (b) not sensitive to colour;
   (c) operate in bright light,
   (d) the receptors giving photopic vision;
   (e) the receptors giving scotopic vision.

2. In the human eye cones are:
   (a) not sensitive to colour;
   (b) sensitive to colour;
   (c) the receptors giving photopic vision;
   (d) the receptors giving scotopic vision;
   (e) the receptors found entirely in the fovea.

3. The peak sensitivity of the human eye occurs:
   (a) at 450 nm; (b) at 555 nm; (c) at 650 nm;
   (d) in the blue part of the spectrum;
   (e) in the red part of the spectrum;
   (f) in the yellow green part of the spectrum.

4. The unit of luminous flux is:
   (a) the candela; (b) the lumen; (c) the lux;
   (d) the candela per square metre.

5. The candela is the unit of
   (a) luminous flux;
   (b) intensity,
   (c) illuminance;
   (d) luminance.

6. The lux is:
   (a) the unit of luminous flux;
   (b) the unit of luminance;
   (c) the unit of illuminance.

7. A uniformly diffusing sphere of diameter 0.2 m has a luminance of 4000 cd/m². Calculate:
   (a) the intensity in any direction (b) the luminous flux emitted.

8. A matt surface has a luminance of 80 cd/m² when the illuminance is 400 lux. Determine the reflectance of the surface.

9. A typical luminous efficacy for an incandescent lamp is:
   (a) 3 lumens/watt;
   (b) 12 lumens/watt;
   (c) 80 lumens/watt;
   (d) 120 lumens/watt.

10 A typical luminous efficacy for a tungsten halogen lamp is:
   (a) 12 lumens/watt;   (c) 80 lumens/watt;
   (b) 25 lumens/watt;   (d) 120 lumens/watt.

11 The luminous efficacy of a fluorescent tube may be up to about:
   (a) 12 lumens/watt;   (c) 120 lumens/watt;
   (b) 80 lumens/watt;   (d) 240 lumens/watt.

12 The spectrum of an incandescent lamp is:
   (a) continuous with emphasis towards the blue end of the spectrum;
   (b) continuous with sharp peaks;
   (c) continuous with emphasis towards the red end of the spectrum;
   (d) a series of discrete lines.

13 The spectrum of a fluorescent lamp is:
   (a) continuous with sharp peaks;   (c) a single spectral line.
   (b) a series of discrete lines;

14 The colour temperature of a filament lamp is:
   (a) about 2700°K;   (c) lower than that of natural daylight;
   (b) about 6000°K;   (d) above that of natural daylight.

15 The colour appearance of a filament lamp would be described as:
   (a) warm; (b) intermediate; (c) cool;
   (d) good but emphasises blue colours.

16 The colour appearance of an artificial daylight fluorescent tube would be described as:
   (a) warm; (b) intermediate; (c) cool;
   (d) good since it can be used for colour matching.

17 The colour rendering of a filament lamp could be described as:
   (a) good but red colours emphasised;
   (b) good but blue colours emphasised;
   (c) poor since most colours are distorted;   (d) warm.

18 The colour rendering of an artificial daylight fluorescent tube could be described as:
   (a) warm   (c) good since it can be used for colour matching
   (b) cool   (d) good but red colours emphasised.

19 In an additive colour system with red, green and blue primaries:
   (a) red plus green gives magenta;   (d) red plus magenta gives yellow;
   (b) red plus green gives yellow;   (e) green plus blue gives cyan;
   (c) yellow plus blue given white;   (f) green plus magenta gives white.

20 In a subtractive colour system with cyan, magenta and yellow primaries:
   (a) cyan plus yellow gives blue;   (d) yellow plus magenta gives green;
   (b) cyan plus yellow gives green;   (e) cyan plus yellow plus magenta gives black;
   (c) yellow plus blue gives black;   (f) red plus green gives yellow.

21 In the Munsell system value describes:
   (a) the actual colour;   (c) the intensity of the colour.
   (b) the degree of whiteness or greyness;

22 A paint has a Munsell colour of 5Y6/2, estimate its reflectance.

11 Lighting calculations

DIRECT COMPONENT OF ILLUMINANCE DUE TO A POINT SOURCE

A point source in practice is such that none of the luminaire dimensions exceed one fifth of the distance between the luminaire and the point under consideration.

Case (i)

The point at which the illuminance is required is directly below the luminaire. The

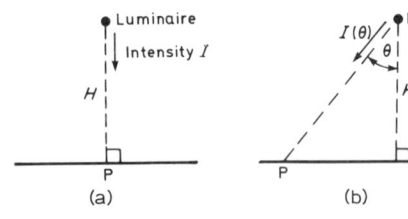

Fig 1(a) Illuminance directly beneath a point source
Fig 1(b) Illuminance at a point not directly beneath a point source

situation is illustrated in *Fig 1(a)*; the illuminance on the horizontal plane, denoted by $E(H)$ is given by $E(H) = \dfrac{I}{H^2}$

where I is the intensity in the vertically downward direction. This is termed the Inverse Square Law.

> *Problem 1* Find the illuminance on a horizontal plane at a point 3 m vertically below a small luminaire which has an intensity of 1700 cd in the vertically downwards direction.

Using $E(H) = \dfrac{I}{H^2} = \dfrac{1700}{3^2} = 188.9$ lux

Case (ii)

If the point P is not directly below the luminaire, as shown in *Fig 1(b)*, then the illuminance on the horizontal plane is given by:

$$E(H) = \dfrac{I(\theta) \cos^3 \theta}{H^2}$$

where $I(\theta)$ is the intensity in the direction of P.

Problem 2 Find the illuminance on the horizontal plane at a point 3 m below and 1.5 m to one side of a luminaire which has an intensity of 1600 cd in the direction of the point.

The situation is shown in *Fig 2* and it is firstly necessary to find the angle θ.

$$\tan \theta = \frac{1.5}{3} = 0.5$$
$$\theta = 26.6°$$

It will be found that $\cos 26.6° = 0.8944$.

Now using: $E(H) = \frac{I(\theta) \cos^3 \theta}{H^2} = \frac{1600 \times (0.8944)^3}{3^2} = 127$ lux

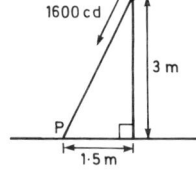

Fig 2 Diagram for Problem 2

Polar curves

The intensity of a source may vary with the angle from which it is viewed. This variation may be illustrated by means of a **polar curve**. *Fig 3* shows a possible polar curve for a luminaire which is symmetric. Note that the intensity is given in candela per 1000 lumens produced by the lamp. Hence the lumen output of the lamp must be known. For non-symmetric luminaires, for example fluorescent luminaires, both axial and transverse polar curves will be required.

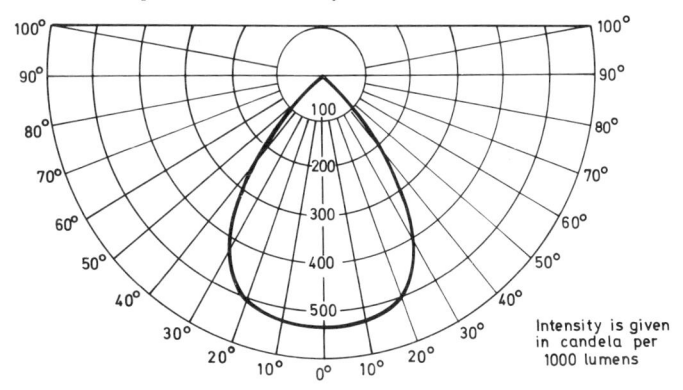

Fig 3 Polar curve for a symmetric luminaire

Intensity is given in candela per 1000 lumens

Problem 3 Find the illuminance on a horizontal plane at a point 3 m below and 1 m to one side of a luminaire which has a polar curve as shown in *Fig 3*. The lamp in the luminaire produces 4000 lumens.

By considering *Fig 4*: $\tan \theta = \frac{1}{3}$

$$\theta = 18.4° \text{ and } \cos \theta = 0.9487$$

From *Fig 3* it can be seen that the intensity at an angle of 18° is approximately 510 candelas/1000 lumens. As the lamp lumens are given as 4000 then the actual intensity is:

$$I(\theta) = \frac{510 \times 4000}{1000} = 2040 \text{ cd}$$

Using the equation for the illuminance on the horizontal plane:
$$E(H) = \frac{I(\theta)\cos^3\theta}{H^2} = \frac{2040 \times (0.9487)^3}{3^2} = 193.5 \text{ lux}$$

Fig 4 Diagram for Problem 3

Case (iii)

The illuminance on a vertical plane, denoted by $E(V)$, for the situation shown in *Fig 5* is given by:

$$E(V) = \frac{I(\theta)\cos^2\theta \sin\theta}{H^2}$$

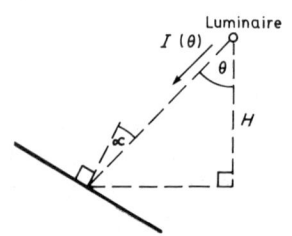

Fig 5 Illuminance at a point on a vertical plane

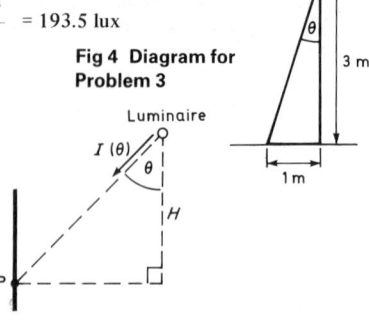

Problem 4 Find the illuminance at a point, P, on a vertical wall due to a luminaire which has a polar curve as shown in *Fig 3*. The lamp in the luminaire produces 4000 lumens. The relevant dimensions are shown in *Fig 6*.

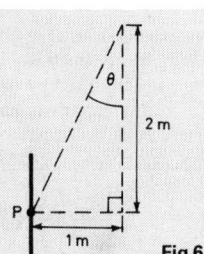

Fig 6

From *Fig 6* it will be seen that:
$$\tan\theta = \frac{1}{2} \text{ hence } \theta = 26.6°$$

It will also be found that: $\cos 26.6° = 0.8944$
and $\sin 26.6° = 0.4472$.

The intensity at an angle of 26° can be read from *Fig 3* as approximately 470 candelas/1000 lumens. Remembering that the lamp produces 4000 lumens the actual intensity is:

$$I(\theta) = \frac{470 \times 4000}{1000} = 1880 \text{ cd}$$

Using the equation for the illuminance on a vertical plane:
$$E(V) = \frac{I(\theta)\cos^2\theta \sin\theta}{H^2} = \frac{1880 \times (0.8944)^2 \times 0.4472}{2^2} = 168 \text{ lux}$$

Case (iv)

For planes which are inclined but are not horizontal or vertical the illuminance, denoted by $E(\alpha)$ is given by:

$$E(\alpha) = \frac{I(\theta)\cos^2\theta \cos\alpha}{H^2}$$

where the angles are illustrated in *Fig 7*.

Fig 7 Illuminance at a point on an inclined plane

Problem 5 Find the illuminance at a point on an inclined drawing table due to a luminaire having a polar curve as shown in *Fig 3*. The lamp lumens are 4000. The relevant dimensions are shown in *Fig 8*.

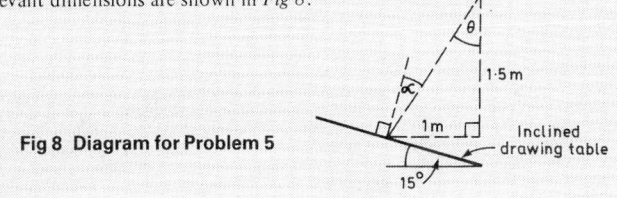

Fig 8 Diagram for Problem 5

It is necessary to determine the angles θ and α as follows:

$\tan \theta = \dfrac{1}{1.5}$; hence $\theta = 33.7°$

and $\alpha = \theta - 15° = 33.7° - 15° = 18.7°$

by considering the geometry of the diagram. From *Fig 3* it will be found that the intensity at an angle of 33.7° is approximately 350 candela per 1000 lumens. As in *Problem 4* the actual intensity is:

$$I(\theta) = \dfrac{350 \times 4000}{1000} = 1400 \text{ cd}$$

The illuminance can now be calculated using:

$$E(\alpha) = \dfrac{I(\theta) \cos^2 \theta \, \cos\alpha}{H^2} = \dfrac{1400 \cos^2 (33.7°) \cos (18.7°)}{1.5^2}$$

$$= 408 \text{ lux}$$

DIRECT COMPONENT OF ILLUMINANCE DUE TO A LINEAR SOURCE

For a linear source such as a long fluorescent tube the illuminance may be found using aspect factors. *Table 1* shows the photometric data for a fluorescent luminaire. The use of this data will be illustrated throughout the remainder of this chapter. Note initially that both axial and transverse polar curves are given, these are drawn on the same graph by drawing one half of each curve. Detailed figures for specifying these curves are also given.

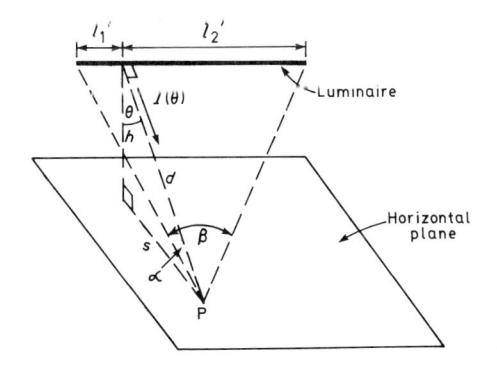

Fig 9 Illuminance on a horizontal plane for a linear source

TABLE 1 Photometric data for fluorescent luminaire
(Reproduced by permission of Thorn lighting)

PPC 675

Description	Single prismatic controller 1800mm
Report Number	500 IL 5295 1
Light Output Ratio	Up .33 Down .46 Total .79
Max. Spacing to Height Ratio (SHR MAX)	1.65

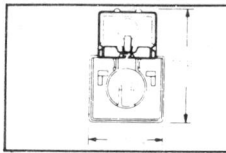

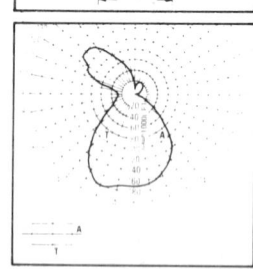

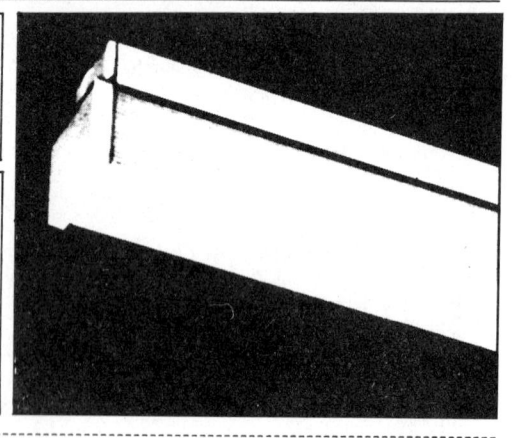

Luminous Intensity in cd/1000lm

Angle (degrees)	Transverse Plane(T)	Axial Plane(A)
0	169	169
5	169	168
10	173	166
15	177	162
20	180	156
25	178	149
30	168	139
35	154	127
40	133	111
45	114	91
50	92	69
55	73	49
60	59	35
65	49	25
70	42	18
75	37	12
80	33	7
85	31	1
90	30	0
95	32	0
100	37	3
105	58	5
110	85	9
115	100	12
120	112	16
125	122	18
130	123	21
135	117	24
140	100	24
145	84	25
150	71	25
155	58	25
160	47	25
165	39	24
170	30	23
175	23	22
180	21	21

Aspect Factors

Angle (degrees)	Parallel Plane	Perpendicular Plane
0	0.000	0.000
5	.087	.004
10	.173	.015
15	.256	.033
20	.334	.058
25	.407	.088
30	.473	.123
35	.532	.160
40	.581	.198
45	.619	.233
50	.647	.263
55	.666	.287
60	.677	.306
65	.684	.319
70	.689	.330
75	.691	.337
80	.692	.342
85	.692	.344
90	.692	.345

Luminance distribution in Cd/m² 1000lm

Angle (degrees)	Transverse Plane(T)	Axial Plane(A)
45	477.8	665.4
50	392.1	555.0
55	318.9	441.7
60	266.5	361.9
65	230.9	305.9
70	208.5	272.1
75	195.6	239.7
80	188.1	208.4
85	193.4	59.3

Utilisation Factors UF[F]   SHR NOM =1.50

Room Reflectances			Room Index								
C	W	F	.75	1.00	1.25	1.50	2.00	2.50	3.00	4.00	5.00
.70	.50	.20	.41	.47	.52	.55	.60	.63	.66	.69	.71
	.30		.36	.42	.47	.50	.56	.59	.62	.66	.68
	.10		.32	.38	.43	.47	.52	.56	.59	.63	.66
.50	.50	.20	.37	.42	.46	.49	.53	.55	.57	.60	.61
	.30		.33	.38	.42	.45	.49	.52	.55	.57	.59
	.10		.29	.34	.39	.42	.47	.50	.52	.56	.58
.30	.50	.20	.33	.37	.40	.43	.46	.48	.49	.51	.53
	.30		.29	.34	.37	.40	.43	.46	.48	.50	.51
	.10		.27	.31	.35	.38	.41	.44	.46	.48	.50
.00	.00	.00	.23	.26	.28	.30	.33	.35	.36	.38	.39
BZ Class			3	4	4	4	4	4	4	4	5

CIE Flux Code 52/ 80/ 92/ 58/ 79

Flux Fraction Ratio .72

Glare Indices

Ceiling reflectance	.70	.70	.50	.50	.30	.70	.70	.50	.50	.30
Wall reflectance	.50	.30	.50	.30	.30	.50	.30	.50	.30	.30
Floor reflectance	.14	.14	.14	.14	.14	.14	.14	.14	.14	.14

Room dimension		Viewed crosswise					Viewed endwise				
X	Y										
2H	2H	7.3	8.5	8.4	9.7	11.1	6.7	8.0	7.8	9.1	10.6
	3H	9.4	10.5	10.5	11.6	13.1	8.4	9.5	9.5	10.7	12.2
	4H	10.4	11.4	11.5	12.6	14.1	9.1	10.2	10.3	11.4	12.9
	6H	11.5	12.5	12.6	13.6	15.2	9.8	10.7	10.9	11.9	13.5
	8H	12.1	13.0	13.3	14.2	15.8	10.0	10.9	11.1	12.1	13.7
	12H	12.8	13.7	13.9	14.9	16.4	10.1	11.0	11.2	12.2	13.7
4H	2H	8.1	9.1	9.2	10.3	11.8	7.6	8.7	8.8	9.8	11.4
	3H	10.5	11.4	11.6	12.6	14.1	9.6	10.5	10.8	11.7	13.2
	4H	11.7	12.5	12.8	13.7	15.3	10.5	11.3	11.7	12.5	14.1
	6H	13.0	13.8	14.2	15.0	16.6	11.4	12.1	12.6	13.3	14.9
	8H	13.8	14.5	15.0	15.7	17.3	11.7	12.4	12.9	13.6	15.2
	12H	14.6	15.2	15.8	16.4	18.1	11.9	12.5	13.1	13.8	15.4
8H	4H	12.3	13.0	13.5	14.2	16.8	11.3	12.0	12.5	13.2	14.8
	6H	13.9	14.5	15.2	15.7	17.4	12.5	13.1	13.7	14.3	15.9
	8H	14.9	15.4	16.1	16.6	18.2	13.0	13.5	14.2	14.7	16.4
	12H	15.9	16.4	17.2	17.6	19.3	13.4	13.8	14.6	15.1	16.7
12H	4H	12.4	13.0	13.6	14.2	15.8	11.6	12.2	12.8	13.4	15.0
	6H	14.1	14.7	15.4	15.9	17.5	12.9	13.4	14.1	14.6	16.3
	8H	15.2	15.7	16.5	16.9	18.6	13.5	14.0	14.7	15.2	16.9
	12H	16.1	16.5	17.4	17.8	19.4	13.7	14.1	15.0	15.4	17.0

Conversion Terms

Luminaire Length (mm)	600	1200	1500	1800	2400
Wattage (W)	1 X 20	1 X 40	1 X 65	1 X 75	1 X 125
Conversion Factors (PC & UF)	1.03	1.09	1.03	1.00	1.05
Glare Indices Conversion	3.82	1.41	.63	0.00	-1.00

The illuminance on a horizontal plane for the simple case shown in *Fig 9* can be calculated using:

$$E(H) = \frac{I(\theta)\,(AF(\alpha) + AF(\beta))\cos\theta}{(l + l'_2)d}$$

where l'_1, l'_2, d and θ are shown in *Fig 9*; $AF(\alpha)$ and $AF(\beta)$ are the aspect factors for a **parallel plane** corresponding to angles α and β in *Fig 9*; $I(\theta)$ is the luminous intensity at the angle θ obtained from the **transverse** polar curve.

Problem 6 The luminaire, the photometric data for which is given in *Table 1*, is 1800 mm long and is fitted with a fluorescent tube producing 5400 lumens. Determine the illuminance on a horizontal plane at a point 2 m below the centre of the luminaire and 1 m to the side as shown in *Fig 10*.

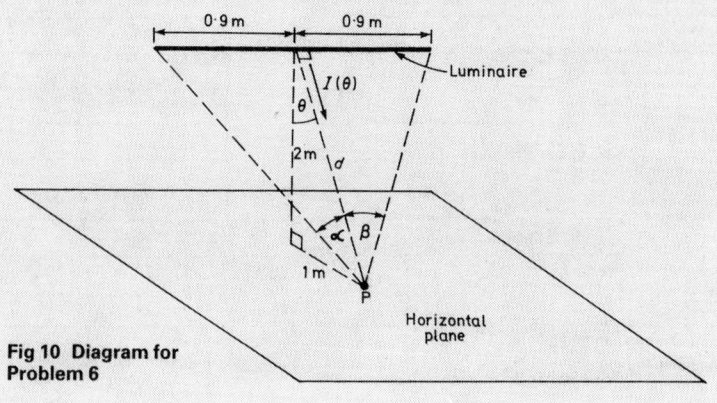

Fig 10 Diagram for Problem 6

Fig 10 shows that $l'_1 = l'_2 = 0.9$ m; $h = 2$ and $s = 1$ m.

It remains to find α, β, θ and d before proceeding. By Pythagoras' theorem:

$$d = \sqrt{(h^2 + s^2)} = \sqrt{(2^2 + 1^2)} = \sqrt{5} = 2.236$$

The angle θ is found as:

$$\tan\theta = \frac{s}{h} = \frac{1}{2}, \text{ whence } \theta = 26.6°$$

In this case the angles α and β are equal and it will be observed that:

$$\tan\alpha = \frac{l'_1}{d} = \frac{0.9}{2.236} = 0.4025 \text{ giving } \alpha = 21.9°$$

Similarly $\beta = 21.9°$

From the data in *Table 1* the luminous intensity at an angle $\theta = 26°$ in the transverse plane is 175 candela per 1000 lumens. As the tube produces 5400 lumens the actual intensity is:

$$I(\theta) = \frac{175 \times 5400}{1000} = 945 \text{ cd}$$

By interpolation of the aspect factors for parallel planes at an angle of $21.9°$ in *Table 1* it will be found that:

$$AF(\alpha) = AF(\beta) = 0.362.$$

Using the formula for the illuminance on a horizontal plane:

$$E(H) = \frac{I(\theta)(AF(\alpha) + AF(\beta)) \cos \theta}{(l'_1 + l'_2)d}$$

$$= \frac{945(0.362 + 0.362) \cos 26.6°}{(0.9 + 0.9) \times 2.236} = 152 \text{ lux}$$

If a point considered is not directly opposite the luminaire then the illuminance on a horizontal plane can be calculated using:

$$E(H) = \frac{I(\theta)(AF(\alpha) - AF(\beta)) \cos \theta}{(l'_1 - l'_2)d}$$

This equation is used in a similar manner to the previous case where α, β, l'_1 and l'_2 are illustrated in *Fig 11*.

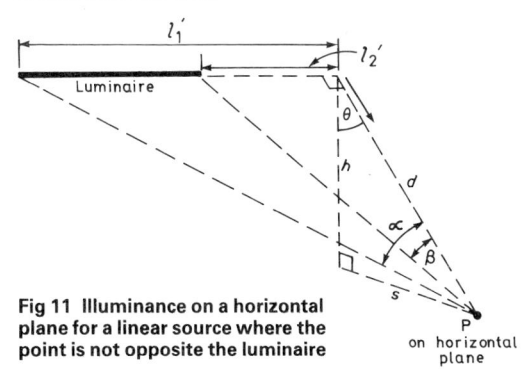

Fig 11 Illuminance on a horizontal plane for a linear source where the point is not opposite the luminaire

LUMEN METHOD

This method permits the calculation of the average illuminance on a horizontal plane due to a regular array of luminaires. The method may be used to design such an array of luminaires for a specified illuminance. It takes into account the light which reaches the horizontal plane both directly from the luminaire and that which reaches it by reflection and inter-reflection from the room surfaces.

The fundamental equation used in this method is:

$$E = \frac{N \times F \times UF \times MF}{A}$$

Where

E = illuminance on a horizontal working plane;

N = number of lamps;

F = lighting design lumens per lamp;

MF = maintenance factor;

UF = utilisation factor, and

A = area of the horizontal surface.

The utilisation factor (UF) is the proportion of the flux produced by the lamps which reaches the working plane. Its value depends on the design of the luminaire, the shape of the room and the effective reflectances of the room surfaces. Values of the utilisation factor are tabulated, in most cases, by the manufacturer of the luminaire.

The maintenance factor (MF) takes account of the loss of light due to dirt accumulating on the luminaires. The value will depend on the cleanliness of the surroundings and the frequency with which the luminaires are cleaned.

The values of the maintenance factor range, typically, from 0.7 to 0.9. The light produced by any lamp decreases with age; for instance the light output of a fluorescent lamp at the end of its life may be as low as 50% of its initial output when new. It is essential that the value of the lighting design lumens for the lamp is used when calculations are made using the maintenance factor.

The important points in this method will be introduced progressively in the following problems and the reader is advised to work though these problems in order.

> *Problem 7* A rectangular room is 10 × 8 m and is 3 m high. The lighting is provided by 16 luminaires in a uniform array having photometric properties as in *Table 1*. The following data may be used:
> Luminaires: 1800 mm long with single 75 w white tubes; fixed to ceiling.
> Lamp output: 5750 lumens (lighting design lumens).
> Room reflectances: ceiling 0.70; walls 0.50; floor 0.20.
> Calculate the average illuminance on the floor, assuming a maintenance factor of 0.85.

The first stage is to calculate the room index, RI, which is a measure of the shape of the room. This is defined as:

$$RI = \frac{L \times W}{(L + W) H}$$

Where
L = length of the room;
W = width of the room;
H = mounting height of the luminaires above the plane at which the illuminance is being calculated.

In this problem:

$$RI = \frac{10 \times 8}{(10 + 8) \times 3} = 1.48$$

Note that H = 3 since the luminaires are mounted on the ceiling and the illuminance on the floor is to be calculated.

Consulting the values of utilisation factors given in *Table 1* for ceiling reflectance $C = 0.70$, wall reflectance $W = 0.50$, floor reflectance $F = 0.20$ and for a room index of 1.5 which is sufficiently close to 1.48 to render interpolation unnecessary it is found that the utilisation factor is 0.55. The illuminance on the floor is then calculated as

$$E = \frac{N \times F \times UF \times MF}{A}$$
$$= \frac{16 \times 5750 \times 0.55 \times 0.85}{80} = 538 \text{ lux}.$$

> *Problem 8* The room and lighting installation have the same properties as the room in *Problem 7* with the exception that the walls, which have a reflectance of 0.5 contain 30 m² of glazing with a reflectance of 0.1. Determine the average illuminance on the floor.

As in *Problem 7* the room index is 1.48 which is very close to 1.5. However before using the table of utilisation factors it is necessary to find the average reflectance of the walls. The average reflectance of a surface can be calculated using:

$$RA(S) = \frac{R(1)\,A(1) + R(2)\,A(2) + \cdots\cdots + R(N)\,A(N)}{A(1) + A(2) + \cdots\cdots + A(N)}$$

where

$RA(S)$ = the average reflectance of the surface;
$R(1), R(2)$ ------ are the reflectances of the parts of the surface having areas $A(1), A(2)$ ------.

Note that $RA(W)$, $RA(C)$, $RA(F)$ will be used to denote the average reflectance of walls, ceiling and floor respectively. In this problem it is found that the total wall area is 108 m² of which 30 m² is glazing leaving 78 m² of wall. The average wall reflectance using the above formula is:

$$RA(W) = \frac{0.5 \times 78 + 0.1 \times 30}{78 + 30} = 0.39$$

Using *Table 1* again and interpolating for $W = 0.39$, the utilisation factor is found to be 0.52. Thus the illuminance is:

$$E = \frac{N \times F \times UF \times MF}{A}$$
$$= \frac{16 \times 5750 \times 0.52 \times 0.85}{80} = \mathbf{508\ lux}$$

Problem 9 The room and lighting installation have the same properties as the room in problem 7 with the exception that the luminaires are suspended 0.4 m below the ceiling. Determine the average illuminance on the floor.

Since the luminaires are suspended below the ceiling a ceiling cavity is created as shown in *Fig 12*. It is necessary to find the effective reflectance, $RE(C)$ of the ceiling cavity. This will take into account the inter-reflections occurring within the cavity. The following procedure will determine the effective reflectance:

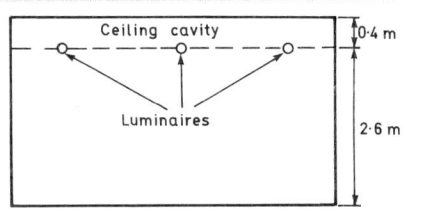

Fig 12 Ceiling cavity in Problem 9

(i) Calculate the average reflectance of the cavity surfaces, $RA(C)$, in a similar manner to that used in *Problem 8*.

(ii) Calculate cavity index, CI, using:

$$CI = \frac{\text{twice mouth area}}{\text{wall area of cavity}}$$

(iii) Calculate $RE(C)$ using:

$$RE(C) = \frac{CI \times RA(C)}{CI + 2(1 - RA(C))}$$

Before proceeding with these calculations the following areas should be noted:
area of ceiling = 80 m² = mouth area of cavity
wall area of cavity = 2(8 × 0.4 + 10 × 0.4) = 14.4 m²

Performing the above procedure:

(i) $RA(C) = \dfrac{R(1)\,A(1) + R(2)\,A(2)}{A(1) + A(2)}$

$$= \frac{0.7 \times 80 + 0.5 \times 14.4}{80 + 14.4} = 0.67$$

(ii) $CI = \frac{2 \times 80}{14.4} = 11.1$

(iii) $RE(C) = \frac{CI \times RA(C)}{CI + 2(1 - RA(C))} = \frac{11.1 \times 0.67}{11.1 + 2(1 - 0.67)} = 0.63$

Since the luminaires are suspended 0.4 m below the ceiling the mounting height, H, is reduced to 2.6 m and the new value of the room index is:

$$RI = \frac{L \times W}{(L + W) H} = \frac{10 \times 8}{(10 + 8) \times 2.6} = 1.7$$

Consulting *Table 1* using the following values $C = 0.63$, $W = 0.50$, $F = 0.20$ and $RI = 1.7$ it will be found by interpolation that the utilisation factor is 0.548 which may be taken as 0.55. The average illuminance on the floor is:

$$E = \frac{N \times F \times UF \times MF}{A} = \frac{16 \times 5750 \times 0.55 \times 0.85}{80} = \mathbf{538 \text{ lux}}.$$

> *Problem 10* The room and lighting installation have the same properties as the room in *Problem 7* with the exception that it is required to calculate the average illuminance on a working plane 0.85 m above the floor

The room index has a new value since the value of H is now 2.15 m:

$$RI = \frac{L \times W}{(L + W) H} = \frac{10 \times 8}{(10 + 8) \times 2.15} = 2.1$$

It will be noted that the utilisation factors in *Table 1* are quoted for a floor reflectance of 0.2. The working plane at a height of 0.85 m creates a floor cavity the effective reflectance of which should be calculated in a manner similar to that used for the ceiling cavity in *Problem 9*. If this effective reflectance differs considerably from 0.2 then a correction will be necessary. The following areas are useful:

Area of floor = 80 m² = mouth area of cavity
Wall area of cavity = 2(8 × 0.85 + 10 × 0.85) = 30.6 m²

The following procedure is that shown in *Problem 9*:

(i) $RA(F) = \frac{R(1) A(1) + R(2) A(2) + \text{---}}{A(1) + A(2) + \text{---}}$

$$= \frac{0.2 \times 80 + 0.5 \times 30.6}{80 + 30.6} = 0.28$$

(ii) $CI = \frac{\text{twice mouth area}}{\text{wall area of cavity}} = \frac{2 \times 80}{30.6} = 5.2$

(iii) $RE(F) = \frac{CI \times RA(F)}{CI + 2(1 - RA(F))} = \frac{5.2 \times 0.28}{5.2 + 2(1 - 0.28)} = 0.22$

This value is very close to 0.2 so that no correction is necessary. Thus from *Table 1* using $C = 0.70$, $W = 0.50$, $F = 0.20$ and $RI = 2.1$ the utilisation factor is 0.60. The average illuminance on the working plane is:

$$E = \frac{N \times F \times UF \times MF}{A} = \frac{16 \times 5750 \times 0.60 \times 0.85}{80} = \mathbf{587 \text{ lux}}$$

Problem 11 It is required to design a lighting system for the room detailed below. The service illuminance required is 500 lux. The luminaires to be used have the properties shown below.

Room properties:
Dimensions: 12 m by 9 m by 3 m high
Height of working plane   0.85 m   Reflectances:
Height of window cills   0.9 m   ceiling   0.7
Height of window heads   2.4 m   wall finishes   0.6
Area of glazing   30 m²   floor   0.2
                                             glazing   0.1

Luminaires Fluorescent as described in *Table 1*, fitted with single 1800 mm 75 W Daylight tube which gives 5400 lumens per tube, suspended 0.5 m below the ceiling. The maintenance factor may be taken as 0.85.

The solution requires that the number of luminaires is determined, a plan of the layout is produced and that the spacing of the luminaires is suitable. A section of the room is

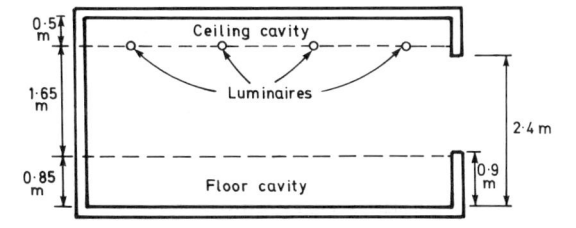

Fig 13 Section of room in Problem 11

shown in *Fig 13* and it will be seen that the glazing is entirely in the walls and not in the ceiling or floor cavities. The following areas arise in the design:

Total walls, excluding walls
of ceiling and floor cavities = 2 × 1.65 × (12 + 9) = 69.3 m²
Windows   = 30 m²
Wall excluding windows   = 69.3 − 30 = 39.3 m²
Wall of ceiling cavity   = 2 × 0.5 × (12 + 9) = 21 m²
Wall of floor cavity   = 2 × 0.85 × (12 + 9) = 35.7 m²

Base of floor and ceiling cavities = 12 × 9 = 108 m²
The average reflectances of the walls, ceiling cavity and floor cavity can now be calculated in the usual manner as illustrated in *Problem 8*.

$$RA(W) = \frac{0.6 \times 39.3 + 0.1 \times 30}{69.3} = 0.38$$

$$RA(C) = \frac{0.6 \times 21 + 0.7 \times 108}{21 + 108} = 0.68$$

$$RA(F) = \frac{0.6 \times 35.7 + 0.2 \times 108}{35.7 + 108} = 0.3$$

The effective reflectances of the ceiling and floor cavities can now be calculated using the method of *Problem 9*.

$$CI \text{ for ceiling cavity} = \frac{\text{twice mouth area}}{\text{wall area of cavity}} = \frac{2 \times 108}{21} = 10.3$$

$$RE(C) = \frac{CI \times RA(C)}{CI + 2(1 - RA(C))} = \frac{10.3 \times 0.68}{10.3 + 2(1 - 0.68)} = 0.64$$

$$CI \text{ for floor cavity} = \frac{\text{twice mouth area}}{\text{wall area of cavity}} = \frac{2 \times 108}{35.7} = 6.05$$

$$RE(F) = \frac{CI \times RA(F)}{CI + 2(1 - RA(F))} = \frac{6.05 \times 0.3}{6.05 + 2(1 - 0.3)} = 0.24$$

The room index can be calculated:

$$RI = \frac{L \times W}{(L + W)H} = \frac{12 \times 9}{(12 + 9) \times 1.65} = 3.1$$

At this stage the following information is available for use in finding the utilisation factor:

$RI = 3.1$, reflectances: ceiling = 0.64, walls = 0.38, floor = 0.24.

The effective reflectance of the floor is sufficiently near 0.2 that the utilisation factor table needs no correction. By interpolation in *Table 1* the utilisation factor is found to be:

$UF = 0.61$

The formula for the illuminance on the horizontal plane can be re-arranged to give the number of lamps required:

$$N = \frac{E \times A}{F \times UF \times MF}, \text{ where } E \text{ is now the specified service illuminance.}$$

$$N = \frac{500 \times 108}{5400 \times 0.61 \times 0.85} = 19.3$$

This suggests that an array of 20 luminaires will be satisfactory.

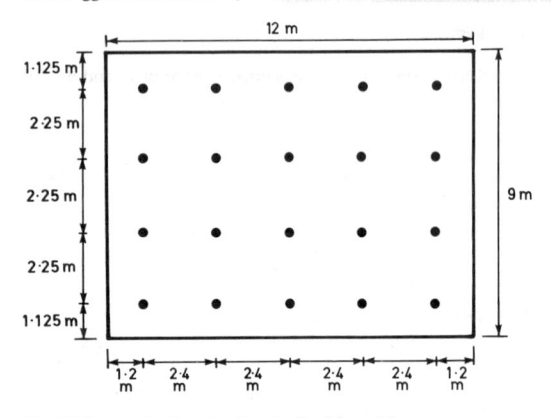

Fig 14 Layout of luminaires in Problem 11

A possible arrangement for the luminaires is shown in *Fig 14*. Note that in the layout, half spacings are used at the walls, for example along the length of the room the spacing of the luminaires is 2.4 m and this is reduced to 1.2 m at the end walls; this is done to avoid dark areas at the walls. It is now essential to ensure that the spacing of the luminaires is suitable.

In order to ensure that the lowest value of the illuminance, occurring between the luminaires, is not less than 70% of the illuminance directly under the luminaires the spacing to height ratio must not exceed SHR MAX given at the top of *Table 1*. The greatest value of the spacing to height ratio for the layout in *Fig 14* is:

$$\frac{\text{spacing}}{\text{height}} = \frac{2.4}{1.65} = 1.45$$

Note that the height used is the distance between the working plane and the luminaire. This value of 1.45 is less than the value of 1.65 for SHR MAX given at the top of *Table 1*, and the spacing is thus satisfactory.

It will be noted that just above the utilisation factors in *Table 1* a value of SHR NOM = 1.50 is given. This is the value of the spacing to height ratio for which the utilisation factors have been calculated. From a practical point of view the utilisation factors are sufficiently accurate for values of the spacing to height ratio within 0.5 of the stated value of SHR NOM. The layout designed above satisfies this criterion since 1.45 is well within 0.5 of SHR NOM = 1.50.

The design given here is necessarily restricted since the photometric data of only one luminaire is given. Using a catalogue of photometric data for a number of luminaires a more economic solution might well be achieved. Such a catalogue would permit the design of a lighting installation for a wide variety of rooms and service illuminances.

GLARE

Good lighting demands that in addition to sufficient illuminance there should be no visual discomfort. Glare can cause discomfort or disability to vision and these two types of glare can be considered separately.

Disability to vision caused by glare reduces the ability to see. The disability glare caused by oncoming car headlights after dark is a well known example. The brightness of the headlamps prevents anything else on the road being seen. Reflections from a shiny surface such as glossy paper or a metal vernier scale will cause disability glare. A large window even with a low brightness sky may produce sufficient intensity to cause disability glare.

Discomfort from glare is distinct from disability due to glare in that the ability to see the visual task is not impaired but the person feels discomfort often in the form of excessive tiredness. In some circumstances both forms of glare can occur simultaneously.

Factors affecting discomfort glare

The factors affecting discomfort glare are:
(i) the brightness of the glaring light source.
(ii) the apparent size of the source at the eye of the observer.
(iii) the position of the source relative to the direction of viewing of the visual task being performed by the observer.
(iv) the general level of brightness of the room surfaces which will be influenced by their reflectances.
(v) the brightness of that part of the walls or ceiling against which the light source is seen. The brighter this immediate background against which the source is seen the less the glare since the contrast between the source and its background is reduced.

GLARE INDEX TABLES

Glare index tables allow a numerical evaluation of discomfort glare for regular arrays of luminaires. The CIBS code for interior lighting gives values for the limiting glare index, this value being the maximum that is likely to be acceptable to the majority of people. In the following problems the use of the glare index tables will be illustrated.

> *Problem 12* A lecture room is lit by a uniform array of luminaires of the type detailed in *Table 1*. The luminaires are 1800 mm long and have single 75 W tubes each producing 5750 lumens. The room is 12 m by 8 m by 3.2 m high. The seating is arranged so that the luminaires are viewed endwise, the direction of viewing being parallel to the length of the room. The reflectances of the room surfaces are: ceiling 0.70, walls 0.50, floor 0.14. Determine the glare index assuming the luminaires to be fixed to the ceiling.

The following stages will illustrate the method:
(i) Determine the height, *H*, of the luminaires above a standardised 1.2 m eye level. This eye level is used since the glare index at eye level. is required and not that at the height of the working plane.
$H = 3.2 - 1.2 = 2$ m.
(ii) Referring to *Fig 15* express the room dimensions as multiples of *H* as found in (i) above. These multiples are denoted as *X* and *Y*. *X* is always the dimension at right angles to the direction of viewing and *Y* is always parallel to the direction of viewing. In this case it will be seen that since $H = 2$ the values of *X* and *Y* are:
$X = 4H$ and $Y = 6H$
(iii) The initial glare index can now be found from *Table 1* using the data:
Ceiling reflectance = 0.70
Wall reflectance = 0.50
Floor reflectance = 0.14
$X = 4H$
$Y = 6H$
Luminaires viewed endwise.

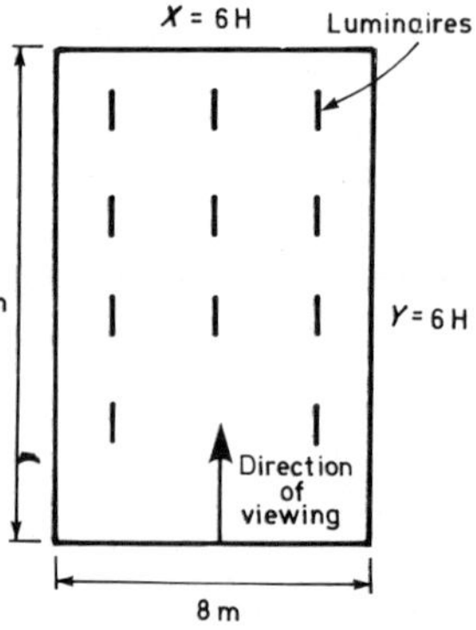

Fig 15 Determination of X and Y in finding a glare index in Problem 12

Inspection of the table gives: initial glare index = 11.4
(iv) It is now necessary to find the conversion term for the downward flux emitted by a luminaire. The light output ratio figures given near the top of *Table 1* show that the proportion of the lamp lumens which are emitted downwards is 0.46. The proportion emitted upwards is 0.33 and thus in total 0.79 of the lamp lumens leave the luminaire. The tube specified for this luminaire produces 5750 lumens thus the downward flux is:
downward flux = 0.46 × 5750 = 2645 lumens.
Consulting *Table 2* it will be found by interpolation that the conversion term for a downward flux of 2645 lumens is:
conversion term = 2.5
(v) A conversion term is also necessary for the height of the luminaire above eye level. The height above eye level is 2 m and it will be found from *Table 2* that the conversion term is −0.5.

TABLE 2 Glare index conversion terms

Downward flux in lumens	Conversion term	Height H above 1.2 m eye level	Conversion term
100	− 6.0	1	− 1.2
150	− 4.9	1.5	− 0.8
200	− 4.2	2	− 0.5
300	− 3.1	2.5	− 0.3
500	− 1.8	3	0.0
700	− 0.9	3.5	+ 0.2
1000	0.0	4	+ 0.4
1500	+ 1.1	5	+ 0.7
2000	+ 1.8	6	+ 1.0
3000	+ 2.9	8	+ 1.4
5000	+ 4.2	10	+ 1.8
7000	+ 5.1	12	+ 2.1
10000	+ 6.0	—	—
15000	+ 7.1	—	—
20000	+ 7.8	—	—
30000	+ 8.9	—	—
50000	+10.2	—	—

(vi) The final glare index is:
 final glare index = initial glare index + conversion terms
  = 11.4 + 2.5 − 0.5 = 13.4

Problem 13 It is required to determine whether a lighting installation will give a final glare index less than a specified limiting glare index of 19. The details are as follows:

Room
Dimensions 12 m by 9 m by 3 m high
Height of working plane 0.85 m
Height of window cills 0.9 m
Height of window heads 2.4 m
Area of glazing 30 m^2

Reflectances:
 ceiling 0.7
 wall finishes 0.6
 glazing 0.1
 floor: assume effective reflectance is 0.14
Usage: as an office with desks arranged so that direction of viewing is at right angles to the 12 m walls.

Luminaires Fluorescent as described in *Table 1*, fitted with single 1800 mm 75 W Daylight tube which produces 5400 lumens; the luminaires are suspended 0.5 m below the ceiling so that direction of viewing is endwise.

The height, H, of the luminaires above a
1.2 m eye level is:
$H = 3 - 1.2 - 0.5 = 1.3$ m. Note that the
luminaires are suspended 0.5 m below
the ceiling.

Expressing the room dimensions in
terms of H it is found that $X = 9.23H$
and $Y = 6.92H$, see *Fig 16*.

Before finding the initial glare index
the effective reflectance of the walls
and ceiling cavity must be found as
shown in *Problem 11*. Since the room
has the same dimensions as that in
Problem 11 the following areas can be
used:

Wall excluding windows = 39.3 m²
Windows = 30 m²
Wall of ceiling cavity = 21 m²
Base of ceiling cavity = 108 m²

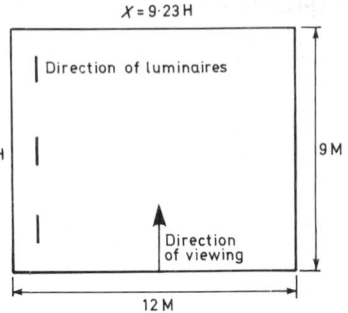

Fig 16 Determination of X and Y in finding a glare index in Problem 13

The average reflectances of the walls and ceiling cavity are:

$$RA(W) = \frac{0.6 \times 39.3 + 0.1 \times 30}{39.3 + 30} = 0.38; \quad RA(C) = \frac{0.6 \times 21 + 0.7 \times 108}{21 + 108} = 0.6$$

The effective reflectance of the ceiling cavity can be calculated as follows:

$$CI \text{ for ceiling cavity} = \frac{\text{twice mouth area}}{\text{wall area of cavity}} = \frac{2 \times 108}{21} = 10.3$$

$$RE(C) = \frac{CI \times RA(C)}{CI + 2(1 - RA(C))} = \frac{10.3 \times 0.68}{10.3 + 2(1 - 0.68)} = 0.64$$

It is now necessary to obtain the initial glare index from *Table 1* using the following
values: $X = 9.23H$, $Y = 6.92H$, ceiling reflectance 0.64, wall reflectance 0.38, floor
reflectance 0.14. This involves considerable interpolation as shown in the following.
The values to be considered from *Table 1*, for endwise viewing, are:

Ceiling reflectance		0.70	0.70	0.50	0.50
Wall reflectance		0.50	0.30	0.50	0.30
Floor reflectance		0.14	0.14	0.14	0.14
X	Y				
8H	6H	12.5	13.1	13.7	14.3
	8H	13.0	13.5	14.2	14.7
12H	6H	12.9	13.4	14.1	14.6
	8H	13.5	14.0	14.7	15.2

Interpolating for wall reflectance of 0.38 gives:

Ceiling reflectance		0.70	0.50
Wall reflectance		0.38	0.38
Floor reflectance		0.14	0.14
X	Y		
8H	6H	12.86	14.06
	8H	13.3	14.5
12H	6H	13.2	14.4
	8H	13.8	15.0

Interpolating for a ceiling reflectance of 0.64 gives:

Ceiling reflectance	0.64
Wall reflectance	0.38
Floor reflectance	0.14

X	Y	
8H	6H	13.22
	8H	13.66
12H	6H	13.56
	8H	14.16

Interpolating for $Y = 6.92H$ gives

Ceiling reflectance	0.64
Wall reflectance	0.38
Floor reflectance	0.14

X	Y	
8H	6.92H	13.42
12H	6.92H	13.84

The final interpolation for $X = 9.23H$ gives

Ceiling reflectance	0.64
Wall reflectance	0.38
Floor reflectance	0.14

X	Y	
9.23H	6.92H	13.55

Thus the initial glare index is 13.55. The conversion terms can now be found. The downward light output ratio given in *Table 1* is 0.46 and the tube produces 5400 lumens thus:

downward flux = 0.46 × 5400 = 2484 lumens.

By interpolation in *Table 2* it will be found that the conversion term for downward flux is 2.33. The mounting height above eye level is 1.3m and again from *Table 2* it will be found that the conversion term is –0.96. The final glare index is:

final glare index = initial glare index + conversion terms
= 13.55 + 2.33 – 0.96
= 14.9

It will be observed that this value is less than the limiting glare index of 19 and the lighting installation will be satisfactory from the point of view of glare.

British zonal system

The above examples of the calculation of glare index were based on the luminaire manufacturers published photometric data. The general method of calculating glare index is described in CIBS Technical Report Number 15. In the report, in order to simplify the calculations, the downward light distribution of luminaires is classified by the BZ (British Zonal) system.

The BZ class number, which ranges from BZ1 to BZ10, classifies the luminaire in terms of the flux directly incident on the working plane relative to the total downward flux. *Fig 17* shows the downward light distribution for three BZ classes. BZ1 has a very concentrated downward light distribution for instance a high bay reflector fitting.

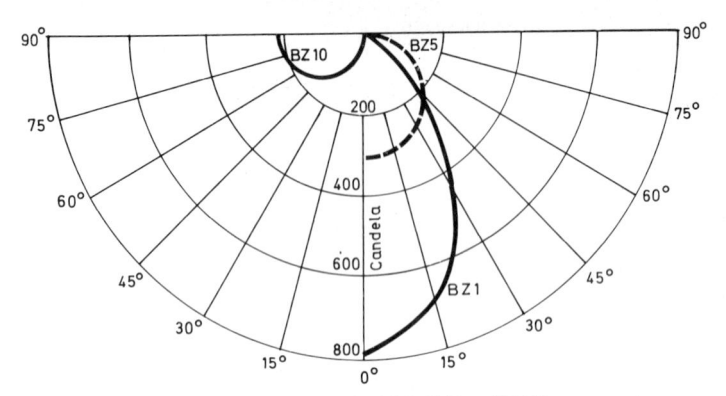

Fig 17 Downward light distributions for BZ1, BZ5 and BZ10

BZ5 is typical of many diffuser fittings. Since the BZ class depends on the flux incident upon the working plane the BZ class can vary with room dimensions.

In *Table 1* it will be seen that the BZ class does change with room index, ranging for this particular luminaire from BZ3 to BZ5. In Technical Report 15 tables of initial glare index are given for each of the BZ classes. Conversion terms are given for endwise or crosswise viewing, high reflectance floor, luminous area, downward flux and height above eye level.

The principle of the use of these tables is exactly similar to that demonstrated in the above examples. In *Table 1* the values for endwise and crosswise viewing are given separately and the conversion term for luminous area has been included. At the foot of *Table 1* conversion terms for different length luminaires are given.

The conversion factors (*PC* and *UF*) apply to the polar curve and utilisation factor and should be used to multiply the downward flux when the conversion term for downward flux is being found. The glare indices conversion should be added to the initial glare index values given in the table. The values in *Table 1* has been calculated for the light distribution of this particular luminaire rather than from the nearest BZ class.

Requirements of good lighting

So far only two criteria have been considered, these are the illuminance on the horizontal plane and the glare index. The lumen method besides ensuring an appropriate value of the average illuminance on the working plane also ensures a suitable uniformity.

Other criteria that may need to be considered at the design stage will be briefly mentioned.
(i) *Colour* The correct selection of colour is obviously important and the CIBS Lighting Code gives recommendations for suitable colour of lamps in many circumstances.
(ii) *Luminance* A proper balance of the luminance between the working surfaces and the walls and ceiling is important in creating an interesting visual environment. The luminance of a surface depends on the illuminance of the surface, due both to direct lighting and light received by reflection from other room surfaces, and the reflectance of the surface.

(iii) *Scalar illuminance* In many rooms where there is a clearly defined visual task the specification of the illuminance on a horizontal plane is not entirely appropriate. The scalar illuminance is the average illuminance on the surface of a small sphere within the space. More simply it is the average illuminance coming from all directions. The CIBS Lighting Code gives a method of converting the scalar illuminance to illuminance on the horizontal plane.

(iv) *Illumination vector* An important influence on the appearance of the human face and other objects is the direction from which the light flows and the strength of the shadows which are cast. The illumination vector can be used to quantify the magnitude and direction of the light flow. The concept of the illumination vector can be envisaged by considering a small sphere at the point in space under consideration.

For some particular diametral plane there will be a maximum difference in illuminance on either side of the plane. The magnitude of this difference gives the magnitude of the illuminance vector and the normal to the diametral plane gives the direction of the illumination vector.

In most situations where the lighting is from a regular overhead layout of luminaires the illumination vector is vertical. During daylight hours natural light from the windows will give a horizontal component to the illumination vector and thus the overall direction of the vector will no longer be vertically downwards. A vector direction of between 15° and 45° below the horizontal produces a pleasing appearance of the human face.

(v) *Vector/scale ratio* The degree of modelling, particularly of the human face, depends on the ratio of the magnitude of the illumination vector to the scalar illuminance. It is suggested that a vector/scalar ratio in the region of 1.2 to 1.8 produces acceptable modelling of the human face particularly if the vector direction is between 15° and 45° below the horizontal. The CIBS Lighting Code illustrates these ideas by a number of photographs.

(vi) *Flicker* Most lamps show a variation of output when operating on 50 Hz alternating current mains as the current reaches a maximum, in one direction or the other, and then goes to zero again. This occurs 100 times a second. Note that for 50 Hz mains the current is zero 100 times in a second.

In filament lamps the variation in filament temperature is small so that the variation in light output is small. For fluorescent lamps and colour corrected mercury lamps the variation is again small since the light emission for the phosphors coating the tube persists for a short while after the current has fallen to zero.

50 Hz flicker occurs in discharge lamps. The polarity of the electrodes alternate 50 times a second and dark spaces in the gas discharge thus change from one end of the tube to the other at this rate. Flicker is more noticeable as the tube ages Electrode shields or special circuits can be used to eliminate flicker.

EXERCISES (answers on page 186)

1   Select the correct option: a point source in practice is such that the distance between the luminaire and the point under consideration is greater than:

   (a) $\frac{1}{5}$ of any luminaire dimension;   (c) twice any luminaire dimension;

   (b) $\frac{1}{2}$ of any luminaire dimension;   (d) 5 times any luminaire dimension.

2. Calculate the illuminance on a horizontal plane at a point 2.5 m vertically below a small luminaire which has a vertically downward intensity of 2000 cd.

3. Find the illuminance on the horizontal plane at a point which is 2 m below and 1.5 m to one side of a luminaire which has an intensity of 1780 cd in the direction of the point.

4. Calculate the illuminance on the horizontal plane at a point 2.0 m below and 1.5 m to one side of a luminaire having the polar curve shown in *Fig 3*. The lamp in the luminaire produces 4500 lumens.

5. Find the illuminance on a vertical plane for the luminaire and point described in *Exercise 4*.

6. Find the illuminance on a plane inclined at 20° to the horizontal for the luminaire and point described in *Exercise 4*.

7. The luminaire, the photometric data for which is given in *Table 1*, is 1800 mm long and is fitted with a fluorescent tube giving 5750 lighting design lumens. Determine the illuminance on a horizontal plane at a point 1.8 m below the centre of the luminaire and 1.2 m to the side in a direction at right angles to the tube.

8. A room is 13 m × 7 m and is 3 m high. The room is lit by a uniform array of 18 luminaires which are fixed to the ceiling. The luminaires, the photometric properties of which are given in *Table 1*, are fitted with fluorescent tubes giving 5400 lighting design lumens. The room reflectances are: ceiling 0.70, walls 0.50, floor 0.20. The maintenance factor is 0.85. Determine the average illuminance on the floor.

9. Repeat *Exercise 8* but assume that the walls contain 35 m^2 of glazing having a reflectance of 0.1.

10. Repeat *Exercise 8* but assume that the luminaires are suspended 0.6 m below the ceiling.

11. Repeat *Exercise 8* but assume that the walls contain 35 m^2 of glazing having a reflectance of 0.1 and that the luminaires are suspended 0.6 m below the ceiling. Assume also that the glazing is entirely in the walls and that no part of it is in the ceiling cavity.

12. It is required to design a lighting system for the room, the plan and section of which are shown in *Fig 18*. The service illuminance is to be 500 lux. The following data is to be used:

Room properties:
Dimensions: 13 m by 7.3 m by 3.1 m
Height of working plane: 0.8 m
Reflectances: ceiling 0.7, wall finishes 0.5, floor 0.2, glazing 0.15.

Luminaires. Fluorescent as in *Table 1* fitted with single 1800 mm tubes which give 5750 lumens per tube. The luminaires are to be suspended 0.5 m below the ceiling and a maintenance factor of 0.8 is considered suitable.
(a) Determine the number of luminaires required.
(b) Check that the proposed spacing is suitable.

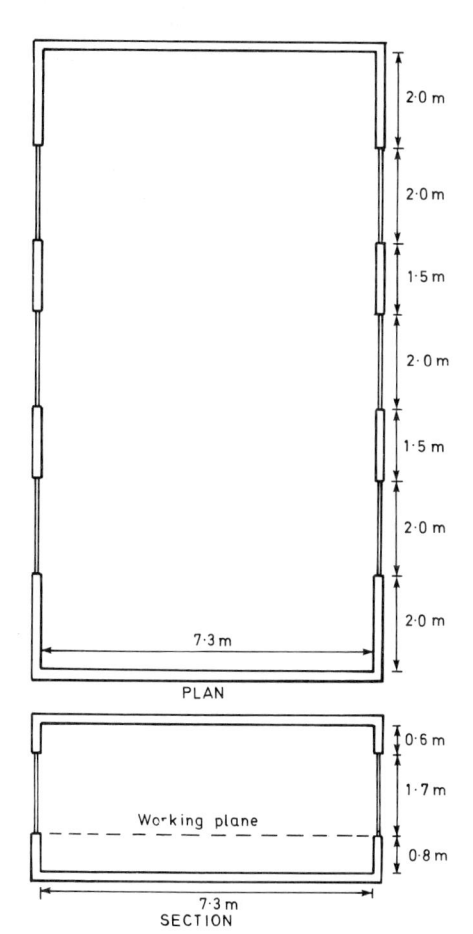

Fig 18 Plan and section of room for Exercise 12

13 A room is lit by a uniform array of luminaires, of the type detailed in *Table 1*, fixed to the ceiling. The luminaires are 1800 mm long and have single tubes giving 5450 lighting design lumens. The room is 12.6 m by 9 m by 3 m high. The effective reflectances are: ceiling 0.7, walls 0.5, floor 0.14. The seating is arranged so that the luminaires are viewed endwise, the direction of viewing being parallel to the length of the room. Determine the glare index.

14 Determine the glare index for the lighting system designed in *Exercise 12*. *The following may be assumed:*
   (i) the glare indices in *Table 1* are applicable;
   (ii) the luminaires are viewed endwise;
   (iii) the direction of viewing is parallel to the longer dimension of the room.

12 Daylighting

THE OVERCAST SKY

The intensity of daylight varies with the latitude, the season, the time of day and the weather conditions. In temperate regions, such as Great Britain there is cloud cover for a considerable amount of time and thus sunlight cannot be relied on for daylighting in buildings. Daylight is at a minimum when the sky is heavily overcast and this condition is used for daylighting calculations.

Two possible types of overcast sky can be used in the prediction of daylight in buildings. The first possible assumption is that the sky has a uniform luminance from horizon to zenith; this is termed a **uniform overcast** sky. Generally the luminance of an overcast sky is not uniform but increases from the horizon to the zenith. For prediction of daylight in buildings the CIE (Commission Internationale de l'Eclairage) **standard overcast** sky is used. For this sky the luminance at an altitude, θ, is given by:

$$L_\theta = L_z \frac{(1 + 2 \sin \theta)}{3}$$

where L_θ is the luminance at altitude θ and L_z is the luminance at the zenith. At the zenith $\theta = 90°$ so that $L_\theta = L_z$. At the horizon $\theta = 0°$ so that $L_\theta = L_z/3$; it will thus be seen that the luminance of the CIE sky at the zenith is three times the luminance at the horizon. Another fact which is of importance in the measurement of daylighting in existing buildings is that the average luminance of the CIE sky is equal to the luminance at an altitude of $42°$.

DAYLIGHT FACTOR AND ITS COMPONENTS

Since daylight is subject to rapid variation it is usual to assess the daylight available in a building by the daylight factor. The daylight factor is defined as:

$$\text{daylight factor} = \frac{\text{daylight illuminance at a point on a given plane}}{\text{simultaneously occurring illuminance outdoors}} \times 100\%$$

Note that the illuminance values to be used are those due to an overcast sky, direct sunlight being excluded. The illuminance value outdoors must be that due to an unobstructed hemisphere of sky.

In order to calculate the daylight factor at a point on a plane in a building it is useful to observe that the daylight reaches the point by three methods.

Fig 1(a) shows light reaching the point under consideration directly from the sky, this is called the **sky component**. *Fig 1(b)* shows light reaching the point after

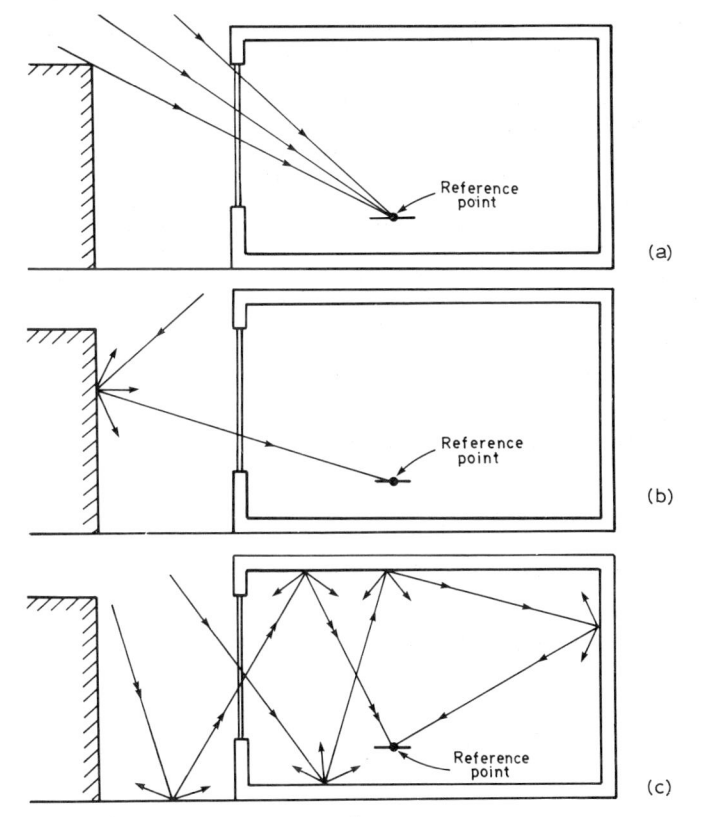

Fig 1(a) Sky component of daylight factor
  (b) externally reflected component of daylight factor
  (c) internally reflected component of daylight factor

reflection from an external obstruction, this is called the **externally reflected component**. Daylight will also reach the point by reflection from other room surfaces as shown in *Fig 1(c)*; this is called the **internally reflected** component. In the prediction of daylight factor it is usual to find the three components separately and then add them together applying any necessary corrections for type of glazing, glazing bars and dirt both on the glazing and the decorations of the room.

PREDICTION OF SKY COMPONENT

Once the luminance distribution for the sky has been assumed it is possible to calculate mathematically the sky component of the daylight factor from the geometry of the windows. Such calculations are not easy and a number of prediction aids have been produced. These include:

(i) The Waldram diagram is a grid representing half the hemisphere of sky and is constructed so that equal areas on the grid represent equal sky components. The outline of the window and any obstructions can be plotted on the grid and the area of unobstructed sky on the grid determined. The sky component is thus determined. The method, although rather slow, will give accurate results particularly where the outline of obstructions is complex.

(ii) The sky dot method which is described in *'Windows and Environment'* published on behalf of the Environmental Advisory Service of Pilkington Flat Glass Ltd. In this method, the window outline and any obstructions are drawn to a suitable scale and an overlay is placed over the drawing. The number of dots on the overlay in the area of unobstructed sky is then counted. Each dot represents $1/10$ per cent sky component.

(iii) BRS simplified daylight factor tables which are published in full in National Building Studies Special Report 26 allow all the components of daylight factor to be found. An introduction to some of these tables will be given later in the chapter. It will be noted that the Building Research Station (BRS) has been renamed Building Research Establishment (BRE) but many publications in common use still have the older abbreviation BRS'

(iv) BRS sky component protractors are a series of ten protractors for finding the sky component for both the uniform overcast sky and the CIE overcast sky for unglazed apertures and glazed apertures. The slopes of glazing covered are: horizontal, vertical, 30° and 60°. The use of a typical protractor will be illustrated later.

BRS SKY COMPONENT TABLE

The BRS sky component table is given in *Table 1* and its use will be illustrated in the following problems.

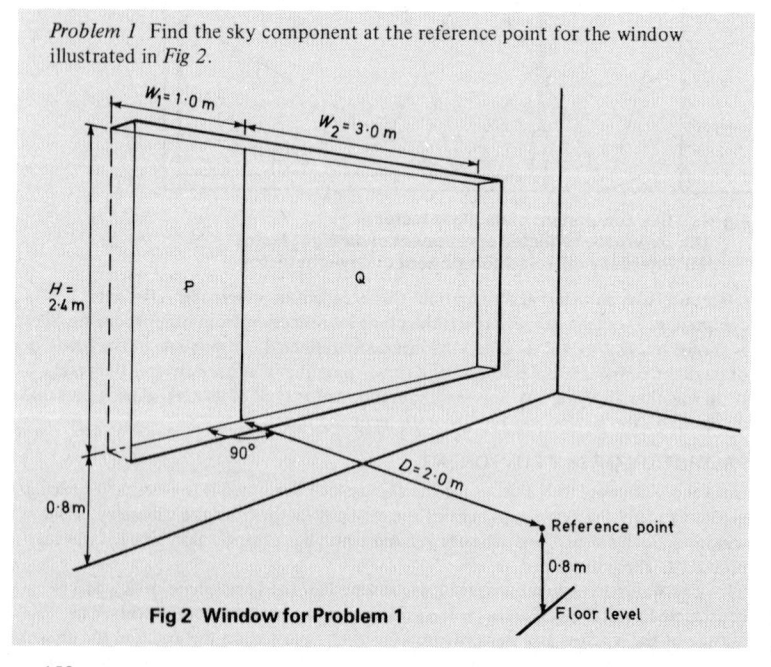

Problem 1 Find the sky component at the reference point for the window illustrated in *Fig 2*.

Fig 2 Window for Problem 1

TABLE 1 Sky components (CIE overcast sky) for vertical glazed rectangular windows

Ratio H/D = Height of window above working plane: distance from window

	0.1	0.2	0.3	0.4	0.5	0.6	0.7	0.8	0.9	1.0	1.1	1.2	1.3	1.4	1.5	1.6	1.7	1.8	1.9	2.0	2.2	2.4	2.6	2.8	3.0	3.5	4.0	5.0	∞
0	0	0	0.1	0.1	0.2	0.2	0.3	0.4	0.5	0.6	0.6	0.7	0.8	0.8	0.9	0.9	0.9	1.0	1.0	1.0	1.1	1.1	1.1	1.1	1.2	1.2	1.2	1.2	1.3
0.1	0	0.1	0.1	0.2	0.4	0.5	0.7	0.8	1.1	1.1	1.3	1.4	1.5	1.6	1.7	1.8	1.9	2.0	2.0	2.1	2.1	2.2	2.2	2.3	2.3	2.4	2.4	2.4	2.5
0.2	0	0.1	0.2	0.3	0.5	0.7	1.0	1.2	1.5	1.7	1.9	2.1	2.3	2.4	2.6	2.7	2.8	2.9	3.0	3.1	3.2	3.3	3.4	3.4	3.5	3.6	3.6	3.7	3.7
0.3	0	0.1	0.3	0.4	0.7	1.0	1.3	1.6	1.9	2.2	2.5	2.7	2.9	3.2	3.3	3.5	3.6	3.8	3.9	4.0	4.1	4.3	4.4	4.5	4.5	4.6	4.7	4.8	4.9
0.4	0	0.1	0.3	0.5	0.8	1.2	1.5	1.9	2.2	2.6	3.0	3.3	3.6	3.8	4.0	4.2	4.4	4.6	4.7	4.8	5.0	5.2	5.3	5.4	5.5	5.7	5.8	5.9	5.9
0.5	0	0.1	0.3	0.6	1.0	1.3	1.7	2.2	2.6	3.0	3.4	3.8	4.1	4.4	4.6	4.9	5.1	5.3	5.4	5.6	5.8	6.0	6.2	6.3	6.4	6.6	6.7	6.8	6.9
0.6	0	0.1	0.3	0.6	1.0	1.3	1.9	2.4	2.8	3.3	3.8	4.2	4.5	4.8	5.1	5.4	5.6	5.8	6.0	6.2	6.4	6.6	6.8	7.0	7.1	7.3	7.4	7.6	7.7
0.8	0.1	0.2	0.4	0.7	1.1	1.6	2.1	2.6	3.1	3.6	4.1	4.5	4.9	5.2	5.6	5.8	6.1	6.3	6.5	6.7	7.0	7.3	7.5	7.6	7.8	8.0	8.2	8.3	8.4
0.9	0.1	0.2	0.4	0.8	1.2	1.7	2.2	2.7	3.3	3.8	4.3	4.8	5.2	5.6	5.9	6.2	6.5	6.7	6.9	7.1	7.4	7.7	7.9	8.1	8.2	8.5	8.7	8.8	9.0
1.0	0.1	0.2	0.4	0.8	1.3	1.8	2.3	2.9	3.4	4.0	4.6	5.0	5.5	5.9	6.2	6.5	6.8	7.1	7.3	7.5	7.9	8.1	8.4	8.6	8.7	9.0	9.2	9.4	9.6
1.2	0.1	0.2	0.5	0.9	1.4	1.9	2.5	3.1	3.7	4.3	4.9	5.4	5.9	6.4	6.8	7.2	7.5	7.8	8.1	8.3	8.7	9.1	9.3	9.6	9.8	10.1	10.3	10.5	10.7
1.4	0.1	0.2	0.5	0.9	1.4	1.9	2.5	3.2	3.8	4.5	5.1	5.7	6.2	6.7	7.1	7.5	7.8	8.2	8.5	8.7	9.1	9.5	9.8	10.0	10.2	10.6	10.9	11.1	11.6
1.6	0.1	0.2	0.5	0.9	1.4	2.0	2.6	3.3	3.9	4.6	5.3	5.9	6.4	7.0	7.4	7.8	8.2	8.5	8.8	9.1	9.6	10.0	10.2	10.5	10.7	11.1	11.4	11.7	12.2
1.8	0.1	0.2	0.5	1.0	1.4	2.0	2.6	3.3	4.0	4.7	5.4	6.0	6.6	7.2	7.6	8.1	8.5	8.8	9.2	9.5	10.0	10.4	10.8	11.1	11.3	11.8	12.0	12.3	12.6
2.0	0.1	0.2	0.5	1.0	1.5	2.0	2.6	3.3	4.0	4.7	5.4	6.1	6.7	7.3	7.8	8.2	8.6	9.0	9.4	9.7	10.2	10.7	11.0	11.4	11.7	12.2	12.4	12.7	13.0
2.5	0.1	0.2	0.5	1.0	1.5	2.1	2.6	3.3	4.0	4.8	5.5	6.2	6.8	7.4	7.9	8.4	8.8	9.2	9.6	9.9	10.5	11.0	11.4	11.7	12.0	12.6	12.9	13.3	13.7
3.0	0.1	0.2	0.5	1.0	1.5	2.1	2.7	3.4	4.1	4.8	5.6	6.2	6.9	7.5	8.0	8.5	8.9	9.3	9.7	10.0	10.7	11.2	11.7	12.0	12.4	12.9	13.3	13.7	14.2
4.0	0.1	0.2	0.5	1.0	1.5	2.1	2.7	3.4	4.1	4.9	5.6	6.3	6.9	7.5	8.0	8.6	9.0	9.4	9.8	10.1	10.8	11.3	11.8	12.2	12.5	13.2	13.5	14.0	14.6
6.0	0.1	0.2	0.5	1.0	1.5	2.1	2.8	3.4	4.2	5.0	5.7	6.3	6.9	7.6	8.1	8.6	9.1	9.5	9.9	10.2	10.9	11.4	11.9	12.3	12.6	13.2	13.6	14.1	14.9
∞	0.1	0.2	0.5	1.0	1.5	2.1	2.8	3.4	4.2	5.0	5.7	6.3	7.0	7.6	8.1	8.6	9.1	9.5	9.9	10.3	10.9	11.5	11.9	12.3	12.7	13.3	13.7	14.2	15.0
0°	6°	11°	17°	22°	27°	31°	35°	39°	42°	45°	48°	50°	52°	54°	56°	58°	60°	61°	62°	63°	66°	67°	69°	70°	72°	74°	76°	79°	90°

Angle of obstruction

(Reproduced from National Building Studies No. 26 by permission of the Controller, HMSO. Crown copyright)

Ratio W/D = Effective width of window to one side of normal: distance from window

TABLE 2

	H/D	W_1/D	W_2/D	Sky component
P	2.4/2 = 1.2	1/2 = 0.5	—	3.3
Q	2.4/2 = 1.2	—	3/2 = 1.5	5.8
			Total	9.1

Note from *Fig 2* that the reference point is at the same height as the window cill. H is the height of the window head above the reference point. W_1 and W_2 are the widths of the window on each side of a line drawn from the reference point normal to the window. D is the distance from the reference point to the outer face of the wall. The sky components for the parts P and Q of the window are found separately and then added. In each case, in order to use *Table 1* the ratios of H/D and W/D must be found. The method is set out in *Table 2*.

The total sky component is 9.1%. The reader is advised to check the values of the sky component in *Table 1* and to note that interpolation was necessary to find the sky component for Q. In this case the reference point is not on the centre line of the window; when this occurs the solution is obviously easier since each half of the window contributes equally to the sky component.

Problem 2 Find the sky component at the reference point for the window illustrated in *Fig 3*.

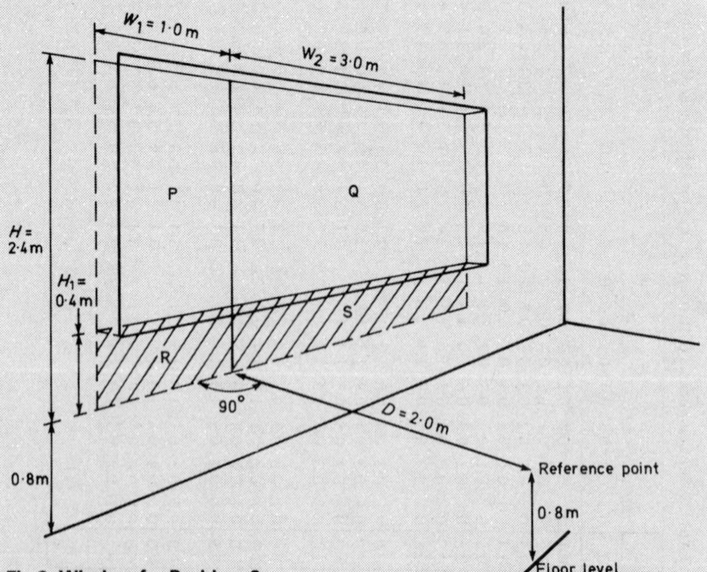

Fig 3 Window for Problem 2

TABLE 3

	H/D	H_1/D	W_1/D	W_2/D	Sky component
PR	2.4/2 = 1.2	—	1/2 = 0.5	—	3.3
QS	2.4/2 = 1.2	—	—	3/2 = 1.5	5.8
R	—	0.4/2 = 0.2	1/2 = 0.5	—	0.1
S	—	0.4/2 = 0.2	—	3/2 = 1.5	0.2

In this case the cill of the window is above the reference point. *Table 1* is constructed such that the values of H to be used are measured from the height of the reference point. Thus sky components must be found for areas PR, QS, R and S separately and the overall sky component is found as PR + QS − R − S. The calculations are laid out in *Table 3*.

Total sky component = 3.3 + 5.8 − 0.1 − 0.2 = **8.8%**

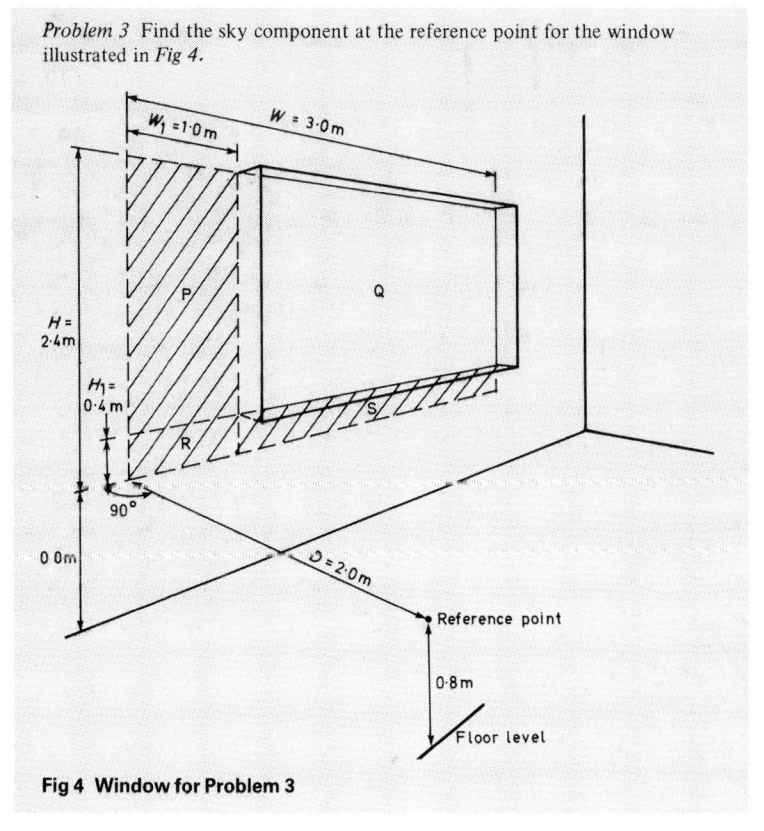

Problem 3 Find the sky component at the reference point for the window illustrated in *Fig 4*.

Fig 4 Window for Problem 3

In this case the reference point is to one side of the window and the cill is above the reference point. The sky components will need to be found for the areas PQRS, PR, RS and R. The overall sky component will be given by PQRS − PR − RS + R. Note that R will have to be added since it has been deducted twice in PR and RS. The calculations are tabulated in *Table 4*.

TABLE 4

	H/D	H_1/D	W/D	W_1/D	Sky component
PQRS	2.4/2 = 1.2	—	3/2 = 1.5	—	5.8
PR	2.4/2 = 1.2	—	—	1/2 = 0.5	3.3
RS	—	0.4/2 = 0.2	3/2 = 1.5	—	0.2
R	—	0.4/2 = 0.2	—	1/2 = 0.5	0.1

Total sky component = 5.8 − 3.3 − 0.2 + 0.1 = **2.4%**

The following additional points in the use of *Table 1* should be noted:

(i) If the reference point is above the window cill that portion of the window below the reference point does not contribute to the sky component and should be ignored in the calculation.

(ii) If deep reveals obscure part of the sky as seen from the reference point then the calculation should be based only on the visible part of the sky. *Fig 5* shows a possible case for a clerestory. The appropriate ratios to be used would be H/D and H_1/D. By similar triangles it will be seen that $H_1/D = H_2/(D − d)$. Generally in such circumstances it would be preferable to use the daylight factor protractors, which will now be described.

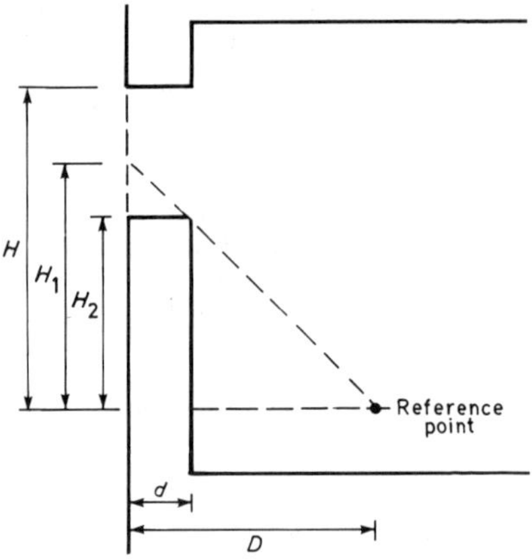

Fig 5 Clerestory for which the cill cuts off some of the direct light

BRS DAYLIGHT PROTRACTORS

The use of these protractors will be illustrated by considering one of the protractors in the series of ten. The one selected will be number 2 in the series which is the sky component protractor for vertical glazing for a CIE overcast sky. This protractor is shown in *Fig 6* and consists of two semicircular sets of scales. The upper scale gives the sky component for infinitely long windows and also has an ordinary angular scale for measuring angles of elevation. The lower scale is used on the plan of the room to obtain corrections for the finite length of the window. The following problems will illustrate the use of the protractor.

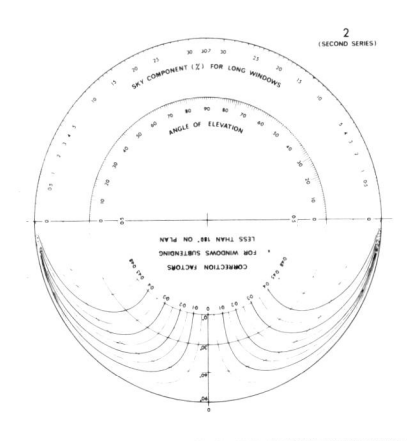

Fig 6 BRS sky component protractor for vertical glazing and a CIE overcast sky (Reproduced by permission of the Controller, HMSO. Crown copyright)

Problem 4 Figs 7(a) and 7(b) show the section and plan of a room for which it is required to find the sky component at the point P on a horizontal reference plane.

Stage 1

The first stage begins on the section of the room; sight lines are drawn from the reference point P to the window cill and window head as shown in *Fig 7(a)*. The protractor is then placed on the section and aligned as shown. Since the visible sky is bounded by the two sight lines the sky component for a long window is the difference of the two readings obtained where the sight lines cross the sky component scale. Reference to *Fig 7(a)* will show that:

Sky component for long window: upper sight line = 2.5%
Sky component for long window: lower sight line = 0.06%

Sky component for long window = 2.44%

At this stage the angles of elevation of the sight lines should be measured on the angle of elevation scale and the average angle of elevation of the visible sky found. Thus:

Angle of elevation: upper sight line = 25°
Angle of elevation: lower sight line = 4°

Average angle of elevation = $\frac{1}{2}(25 + 4) = 14.5°$

Stage 2

The second stage is performed upon the plan of the room; sight lines are again drawn from the reference point P to define the visible area of sky as shown in *Fig 7(b)*. The protractor is then placed on the plan and aligned as shown; ensuring that the zero line is perpendicular to the glazing. Note that on the correction factor scale there are four semicircles marked 0°, 30°, 60° and 90°. These angles refer to the average angle of elevation of the visible sky as found from the section of the room.

In this problem the average angle of elevation has already been found to be 14.5°. A part circle corresponding to this average angle of elevation has been inserted as a

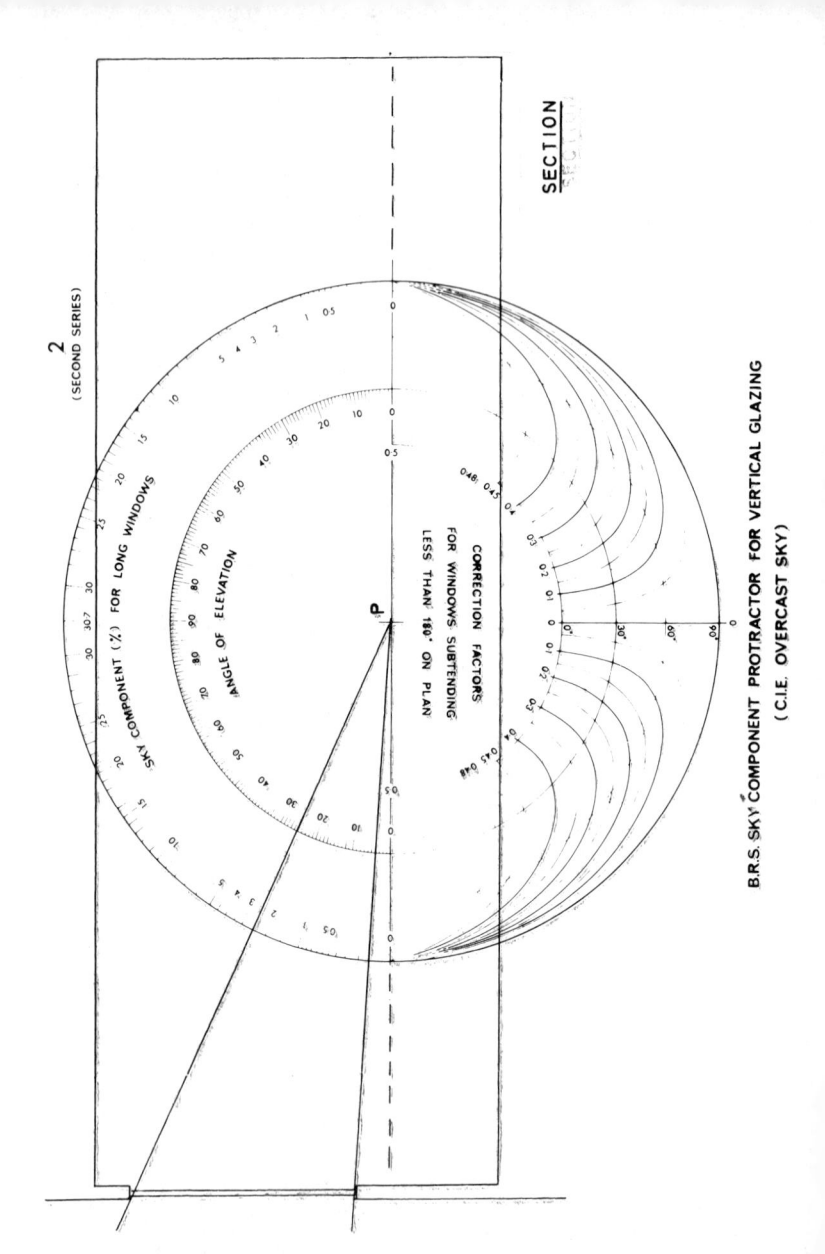

Fig 7(a) section of room and BRS sky component protractor for Problem 4

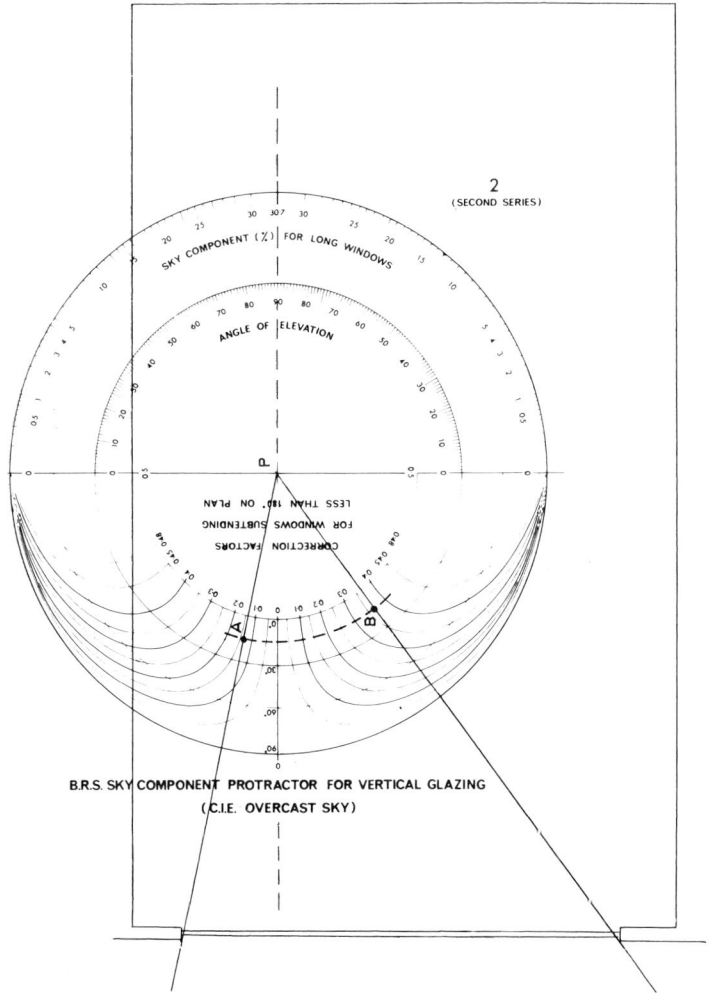

**Fig 7 (b) plan of room and BRS sky component protractor for Problem 4
(BRS protractor reproduced by permission of the Controller, HMSO.
Crown copyright)**

dotted curve on *Fig 7(b)*. The required corrections are found where the lines of sight cross the dotted circle corresponding to the average angle of elevation of 14.5°, these points are denoted by A and B on *Fig 7(b)*.

The numerical values are obtained by estimation between the curved correction factor lines. For instance the point B is about half way between the curves for 0.35 and 0.4 and may thus be estimated as about 0.37. Similarly the correction factor at point A can be estimated as 0.14. Since these correction factors are on opposite sides fo the zero line they are to be added together. If the reference point is to one side of the window the correction factors would be subtracted. In this case:

correction factor for window subtending less than 180° on plan = 0.37 + 0.14 = 0.51
This correction factor is used to multiply the sky component for the long window found above giving the final sky component; thus:

sky component = 0.51 × 2.44 = **1.24%**

PREDICTION OF THE EXTERNALLY REFLECTED COMPONENT

The externally reflected component can be calculated by finding the equivalent sky component for the area of sky obscured by the external obstruction which would otherwise be visible from the reference point and then modifying this value by the luminance of the obstruction. For normal situations the following assumptions are satisfactory:

(i) for a uniform overcast sky the externally reflected component is given by:

$$\frac{\text{equivalent sky component}}{10} \%$$

(ii) for a CIE overcast sky the externally reflected component, for obstructions near horizon level is given by:

$$\frac{\text{equivalent sky component}}{5} \%$$

The different values arise since the luminance of a CIE sky is lower at the horizon than the zenith.

The equivalent sky component can be found by treating the obstruction as if it were a patch of sky and using any of the methods previously described for finding the sky component. The use of the BRS sky component table will be illustrated in the next problem and the use of the BRS protractor in *Problem 8*.

> *Problem 5* Find the externally reflected component for the window illustrated in *Figs 8(a) and 8(b)*. The external obstruction is another building which is continuous across the window and parallel to it. The obstruction has a height of 6 m above the reference point and is 16 m from the outside of the wall containing the window.

The first part of this problem is to find H_1 as shown in *Fig 8(a)*. Several methods are possible and the following is a possibility. In *Fig 8(b)*, by similar triangles:

$$\frac{H_1}{H_2} = \frac{D}{D_1} \text{ giving}$$

$$\frac{H_1}{6} = \frac{2}{18}$$

$$H_1 = 0.67 \text{ m}$$

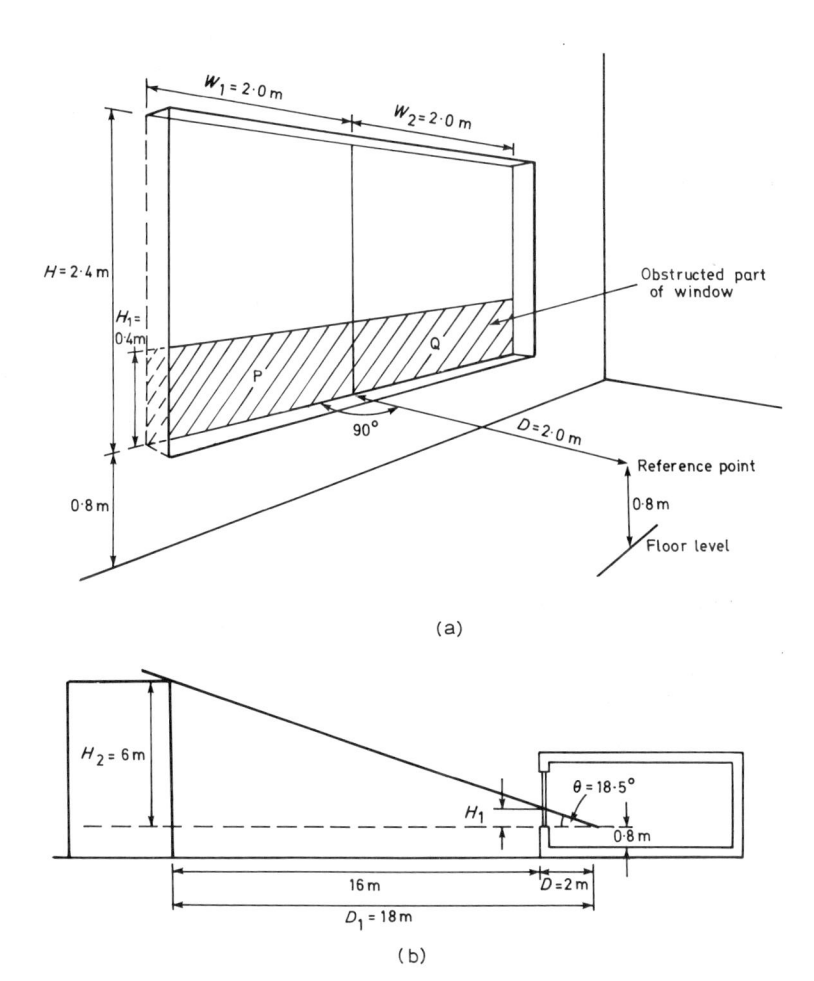

Fig 8(a) Window for Problem 5
(b) Section for Problem 5

TABLE 5

	H_1/D	W_1/D	W_2/D	Sky component
P	0.67/2 = 0.33	2/2 = 1	—	0.53
Q	0.67/2 = 0.33	—	2/2 = 1	0.53
			Total	**1.06**

Alternatively the angle θ, as shown in *Fig 8(b)* may be calculated or obtained from scale drawings as $18.5°$. The calculations may now be set out as a table (see *Table 5*). The values of the sky component were found from *Table 1*; note that the bottom line of the table gives the angle of obstruction and this is useful if the angle of obstruction from the reference point is known, in this case the angle of obstruction is $\theta = 18.5°$.

The equivalent sky component is 1.06 and since this applies for a CIE overcast sky, the externally reflected component is:

$$\text{externally reflected component} = \frac{1.06}{5} = 0.21\%$$

If the outline of the obstruction is irregular it is often possible to draw an equivalent horizontal obstruction line so that the area of obstruction above the line is equal to the area of the sky below it. If an obstructing building does not obstruct the complete width of the window it may be easier to evaluate the two parts separately as if they were two separate windows; see *Problem 8*.

INTERNALLY REFLECTED COMPONENT

The internally reflected component depends on the reflectances of the room surfaces and the amount of light they receive from the sky, the external obstructions and the ground outside. The internally reflected component varies according to the position of the reference point within the room being least at points furthest away from the window

The average value of the internally reflected component may be taken to apply over the majority of the room and the minimum value to apply at points remote from the window. For compliance with statutory requirements the minimum value should be used.

Several possible methods exist for the prediction of the internally reflected component. Amongst these are:
(i) the BRS inter-reflection formula.
(ii) BRS nomograms which are based on a modified form of the BRS inter-reflection formula.
(iii) BRS tables which rapidly predict the minimum internally reflected component under certain limitations concerning the size of room and the surface reflectances. These tables which are simple to use are described in BRE digest No. 42.

BRS inter-reflection formula

This applies to side-lit rooms and is stated as:

$$\text{Average internally reflected component} = \frac{0.85W}{A(1-R)} \times (CR(FW) + 5R(CW))\%$$

where
W = area of window;
A = total surface area of the room (including window);
R = average reflectance of all room surfaces (including window);
$R(FW)$ = average reflectance of surfaces below the plane of the mid-height of the window (excluding the window wall)
$R(CW)$ = average reflectance of surfaces above the plane of the mid-height of the window (excluding the window wall)
C = a coefficient dependant on the angle of obstruction as measured from the centre of the window.

Some values of C are given in *Table 6* for a CIE overcast sky. Further values and assumptions are given in BRE digest No. 42 (see *Table 6* below)

TABLE 6

Angle of obstruction	C
No obstruction	39
10°	35
20°	31
30°	25

(Reproduced from BRE Digest 42 by permission of the Controller, HMSO. Crown copyright)

> *Problem 6* Determine the average internally reflected component of the daylight factor for the room the plan and section of which are shown in *Fig 9*. The following reflectances are to be used:
> Walls 0.5; ceiling 0.7; floor 0.2; window glass 0.15.

In the first instance all necessary areas will be found:
Window area, W $= 1.9 \times 7 = 13.3$ m²
Total surface area of room, A $= 2(9 \times 6 + 9 \times 3 + 6 \times 3) = 198$ m²
Floor area $= 9 \times 6 = 54$ m²
Ceiling area $= 9 \times 6 = 54$ m²
Area of walls (excluding window) $= 2(9 \times 3 + 6 \times 3) - 13.3 = 76.7$ m²
Area of walls above mid-height of window (excluding window wall)
$= 6 \times 1.2 + 9 \times 1.2 + 6 \times 1.2 = 25.2$ m²
Area of wall below mid-height of window (excluding window wall)
$= 6 \times 1.8 + 9 \times 1.8 + 6 \times 1.8 = 37.8$ m²

The average reflectances can now be found using:

$$\text{average reflectance} = \frac{R(1)A(1) + R(2)A(2) + R(3)A(3) + \cdots}{A(1) + A(2) + A(3) + \cdots}$$

where $R(1), R(2) \cdots$ etc. are the reflectances of surfaces having areas $A(1), A(2) \cdots$ etc. This method has already been used in chapter 11. The average reflectance of all room surfaces is:

$$R = \frac{0.5 \times \text{wall area} + 0.7 \times \text{ceiling area} + 0.2 \times \text{floor area} + 0.15 \times \text{window area}}{\text{total surface area}}$$

$$= \frac{0.5 \times 76.7 + 0.7 \times 54 + 0.2 \times 54 + 0.15 \times 13.3}{198} = 0.45$$

The average reflectance of surfaces below the mid height of the window is:

$$R(FW) = \frac{0.5 \times \text{area of walls below mid-height} + 0.2 \times \text{floor area}}{\text{area of walls below mid-height} + \text{floor area}}$$

$$= \frac{0.5 \times 37.8 + 0.2 \times 54}{37.8 + 54} = 0.32$$

Similarly using the area of walls above the mid-height of the window and the ceiling area:

$$R(CW) = \frac{0.5 \times 25.2 + 0.7 \times 54}{25.2 + 54} = 0.64$$

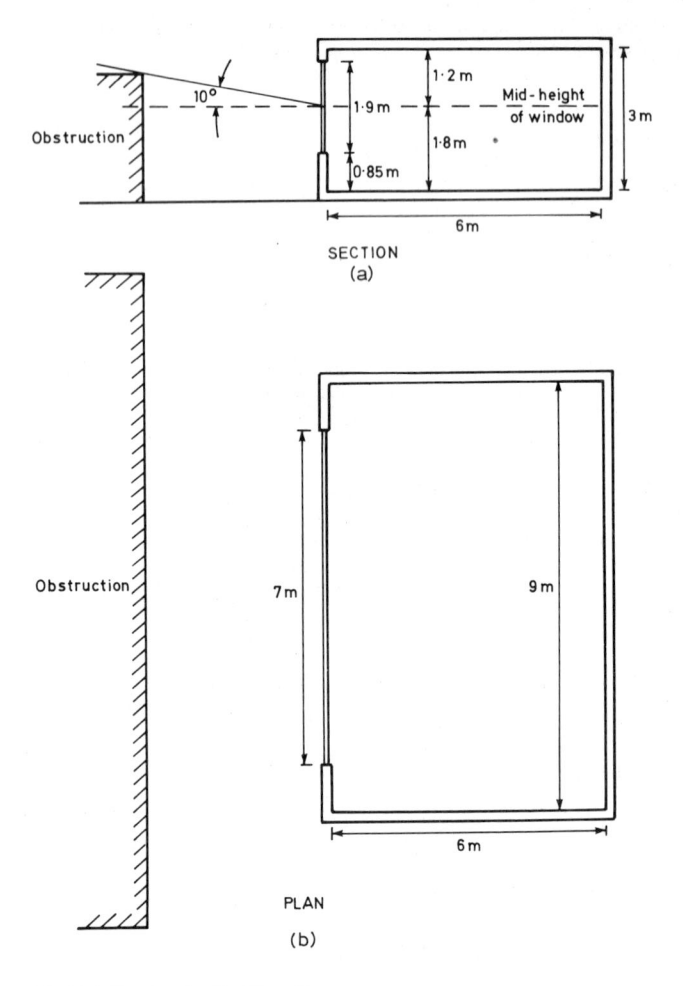

Fig 9(a) Section for Problem 6
(b) Plan for Problem 6

Noting that the value of C from *Table 6* for a 10° obstruction is 35, the average internally reflected component can now be calculated:

$$\text{average internally reflected component} = \frac{0.85W}{A(1-R)} \times (CR(FW) + 5R(CW))\%$$

$$= \frac{0.85 \times 13.3}{198(1-0.45)} \times (35 \times 0.32 + 5 \times 0.64)\%$$

$$= \mathbf{1.5\%}$$

BRS nomograms

These provide a more rapid method than the inter-reflection formula for predicting the internally reflected component. Three nomograms exist and the principle of their use will be shown by considering Nomogram 1. This nomogram predicts the average internally reflected component for a side-lit room, and assumes ceiling and floor reflectances of 0.7 and 0.15 respectively and a ratio of window area to total surface area of 0.05.

> *Problem 7* Repeat *Problem 6* using the BRS Nomogram 1.

Fig 10 shows nomogram 1 and it will be seen that there are five scales lettered A to E. In using the nomogram it is first necessary to determine values for scales A and B.

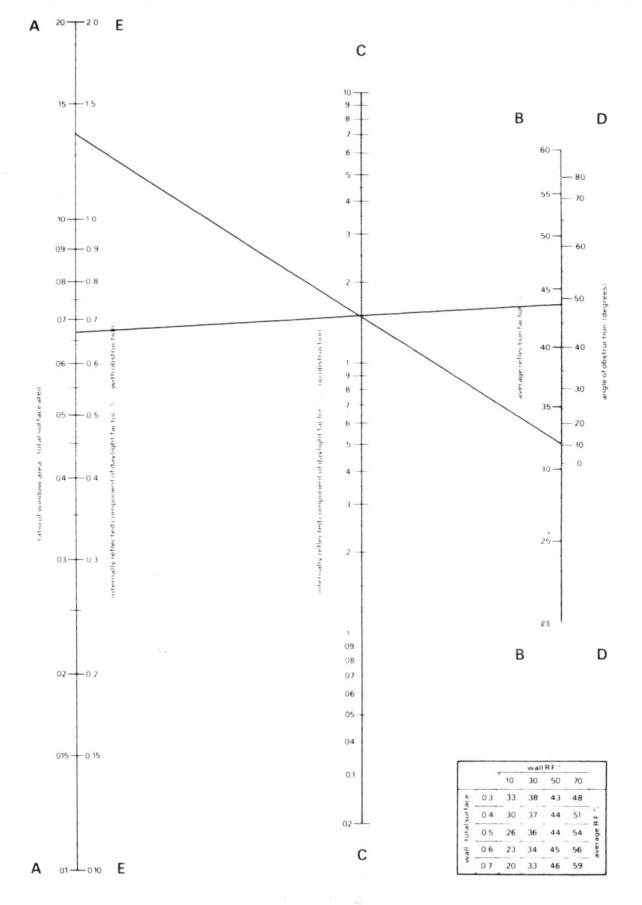

Fig 10 BRS Nomogram 1 for Problem 7 (Reproduced from BRE Digest 42 by permission of the Controller, HMSO. Crown copyright)

For scale A: from *Problem 6* it will be found that:
 Window area   = 13.3 m²
 Total surface area   = 198 m²
Thus
$$\frac{\text{window area}}{\text{total surface area}} = \frac{13.3}{198} = 0.067$$

For scale B: the average reflection factor (*RF*) can be found using the inset table. To use this table note that the reflectance of the walls in *Problem 6* is 0.5 or 50%. The ratio of wall area to total surface area is:
$$\frac{\text{wall area (including window)}}{\text{total surface area}} = \frac{90}{198} = 0.45$$

Consulting the inset table it will be seen that the average reflection factor is 44%. Note that the term reflection factor is synonomous with the term reflectance, expressed as a percentage.

The value 0.067 is marked on scale A and the value 44% is marked on scale B and these points are joined by a straight line. In *Fig 10* it will be seen that this line intersects scale C at a value of 1.5 which would be the value of the internally reflected component if there were no external obstruction. From the section of the room, shown in *Fig 9*, the external obstruction is at an angle of 10° as measured from the mid-height of the window. This value of 10° is plotted on scale D. A line is drawn from this value on scale D through the previously determined value on scale C to intersect scale E. This value gives the internally reflected component with obstruction which is seen to be 1.4%.

This value is slightly lower than that obtained using the inter-reflection formula since the actual floor reflectance is 0.2 whereas the inset table on the nomogram was constructed for a floor reflectance of 0.15.

CALCULATION OF TOTAL DAYLIGHT FACTOR

The total daylight factor is found by summing the three components and applying suitable correction factors. The correction factors are:
(i) Glazing type The methods used to find the components of daylight factor assume clear single glazing up to 6 mm thick. For other glasses a correction factor is necessary. Values of the correction factor can be found in BRE Digest 42 or in manufacturers data.
(ii) *Dirt on glazing* This correction will depend on the location of the building, the nature of work conducted in the building and the inclination of the glazing. Values of this correction factor may be found in BRE Digest 42 or the explanatory book of the BRS daylight protractors.
(iii) *Allowance for glazing bars and window framing* This can be calculated as:
$$\frac{\text{actual glass area}}{\text{area of window aperture}}$$
or typical figures given in the explanatory book of the BRS daylight protractors can be used.
(iv) *Deterioration of decorations* The correction factor applies to the internally reflected component only and suitable values are given in BRE Digest 42.

The calculation of a total daylight factor will be illustrated by a full worked example.

Problem 8 It is required to find the daylight factor at the point P in the room, the plan and section of which are illustrated in *Fig 11*. The reflectance of the wall finishes may be taken as 0.5. The following correction factors may be regarded as appropriate:

Glazing type	1.0
Dirt on glazing	0.9
Glazing bars	0.8
Deterioration of decorations	0.85

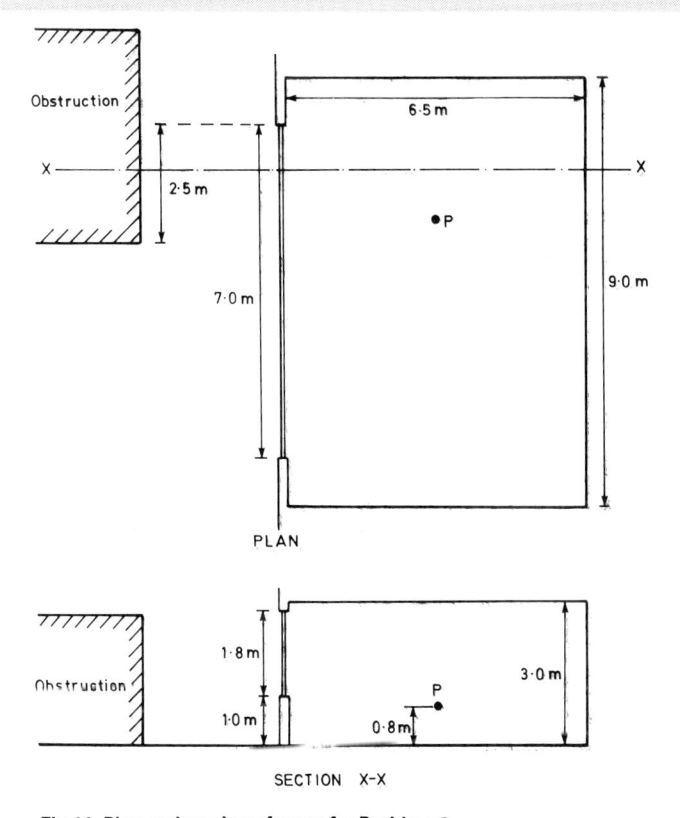

Fig 11 Plan and section of room for Problem 8

The solution will be achieved using the BRS protractor and nomogram. Since the window is only partially obstructed on plan it will be treated as two separate parts: the unobstructed and obstructed part.

(i) Sky component

The sight lines PA, PB and PC are drawn on the section shown in *Fig 12(a)*. Note that the lines PA and PB define the patch of visible sky for the obstructed part of the window, and that the lines PA and PC define the patch of visible sky for the unobstructed part of the windows. On the plan shown in *Fig 12(b)* the sight lines PD and PE define the obstructed part of the sky and the sight lines PE and PF define the unobstructed part of the sky. The sky components are now found in a similar manner to that shown in *Problem 4*.

	Obstructed	Unobstructed
Sky component for long windows: (see *Fig 12(a)*)	3.9 1.0 ――― 2.9	3.9 0.06 ――― 3.84
Average angle of elevation: (see *Fig 12(a)*)	$\frac{30 + 17}{2} = 23.5°$	$\frac{30 + 4}{2} = 17°$
Correction factors for window less than 180° on plan (see *Fig 12(b)* and notes below)	0.3 0.055 ――― 0.355	0.47 0.06 ――― 0.41
Sky component:	2.9 × 0.355 = 1.03	3.84 × 0.41 = 1.57

Total sky component = 1.03 + 1.57 = 2.60

Notes. Since the average angle of elevation of the visible sky is different for the unobstructed and obstructed parts different interpolation circles are required when finding the correction factors. For the obstructed part where the average angle of elevation is 23.5° the required values are obtained at points G and H in *Fig 12(b)*. Since these points are on opposite sides of the zero line the values are *added*. For the unobstructed part the average angle of elevation is 17° and the required values are obtained at the points J and K. Since these are on the same side of the zero line the values are *subtracted*.

(ii) Externally reflected component

The equivalent sky component is found for the obstruction itself as follows.

Sky component for long windows: (see *Fig 12(a)*)	1.0 0.06 ――― 0.94
Average angle of elevation: (see *Fig 12(a)*)	$\frac{17 + 4}{2} = 10.5°$
Correction factors for window less than 180° on plan (see *Fig 12(b)* and notes below)	0.31 0.06 ――― 0.37
Equivalent sky component:	0.94 × 0.37 = 0.35
Externally reflected component:	$\frac{0.35}{5} = 0.07\%$

Notes The obstruction is defined by the sight lines PB and PC on *Fig 12(a)* and has an average angle of elevation of 10.5°. On *Fig 12(b)* this value is used to give the interpolation circle when finding the correction factors. The required values are at the

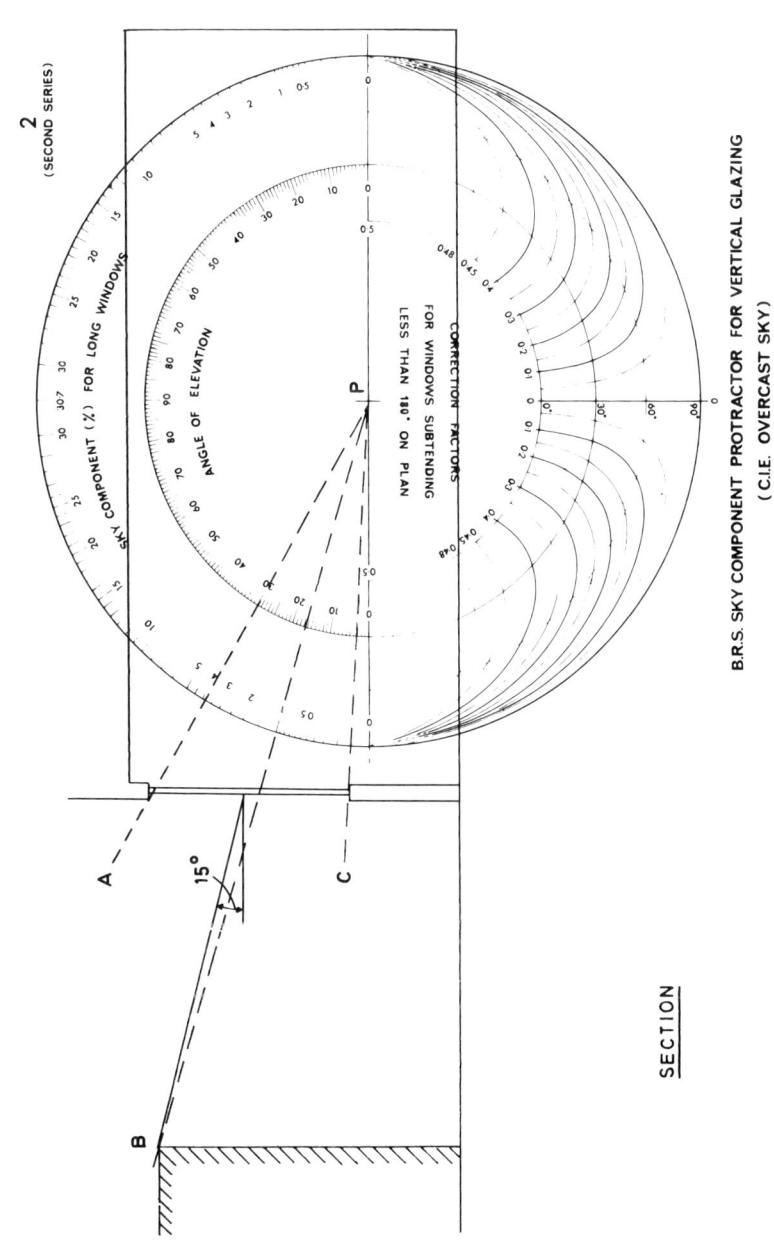

Fig 12(a) Section of room and BRS sky component protractor for Problem 8

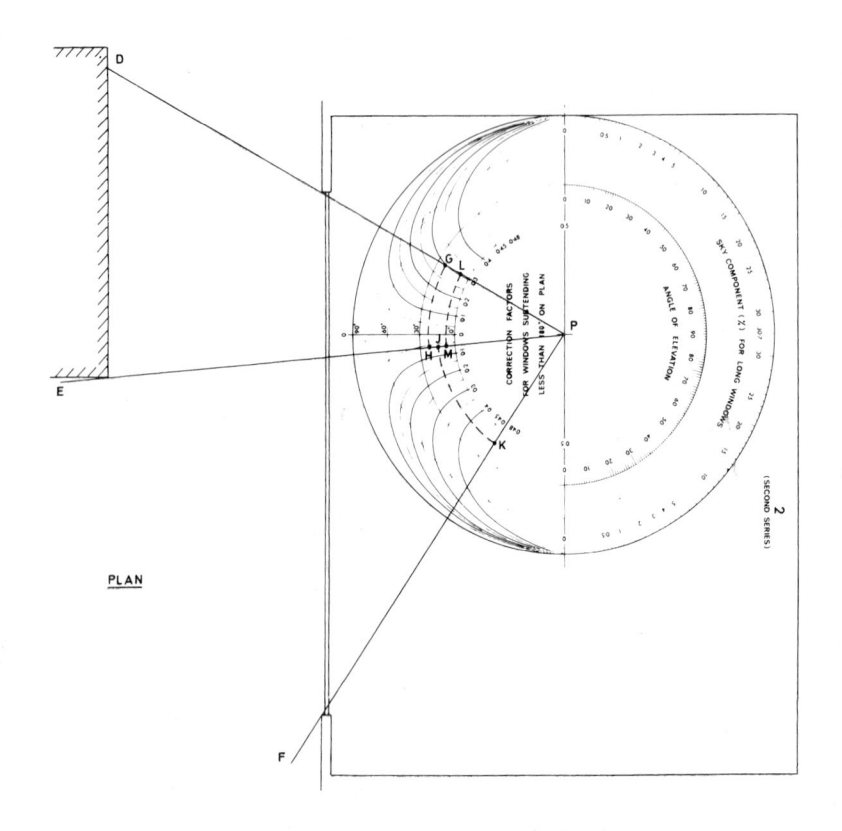

(b) **Plan of room and BRS sky component protractor for Problem 8**
   (BRS protractor reproduced by permission of the Controller, HMSO.
Crown copyright)

points L and M; these values have to be added since they are on opposite sides of the zero line. As previously explained the externally reflected component is taken as one-fifth of the equivalent sky component for a CIE overcast sky.

(iii) *internally reflected component:* the two parts of the window are again treated separately. The following areas will be used:

area of obstructed part of window $= 2.5 \times 1.8 = 4.5$ m^2
area of unobstructed part of window $= 4.5 \times 1.8 = 8.1$ m^2
total surface area of room $= 2(9 \times 6.5 + 9 \times 3 + 6.5 \times 3)$
$= 210$ m^2
area of walls (including window) $= 2(9 \times 3 + 6.5 \times 3) = 93$ m^2

The following table sets out the use of nomogram 1 for both parts of the window.

	Obstructed	*Unobstructed*
Scale A: $\dfrac{\text{window area}}{\text{total surface area}}$	$\dfrac{4.5}{210} = 0.0214$	$\dfrac{8.1}{210} = 0.0386$
Scale B: $\dfrac{\text{wall area}}{\text{total surface area}}$	$\dfrac{93}{210} = 0.44$	$\dfrac{93}{210} = 0.44$
Use inset table on *Fig 13* and given wall RF = 50% to give average reflection factor:	44%	44%
Join A and B as in *Fig 13* to give value on scale C	(0.48)	0.9 – final value since no obstruction
Scale D: angle of obstruction measured from mid-height of window (see *Fig 12(a)*).	15°	
Join C and D (*Fig 13*) and find internally reflected component on Scale E	0.41	

Total internally reflected component = 0.41 + 0.9 = 1.31

(iv) *total daylight* factor: the following values have so far been obtained:

Sky component 2.60
Externally reflected component 0.07
Internally reflected component 1.31

All that now remains to be done is to apply the correction factors; the correction factor for the deterioration of the decorations applies only to the internally reflected component and thus a value of 0.85 × 1.31 will be used for this component. The other three corrections apply to all the components; thus the daylight factor is the sum of the three components multiplied by the correction factors for glazing type, dirt on glazing and glazing bars. Thus:

daylight factor = 1.0 × 0.9 × 0.8 (2.60 + 0.07 + (0.85 × 1.31))
= 0.72 (2.60 + 0.07 + 0.935)
= 0.72 × 3.605
= 2.6%

Notes on Problem 8

This problem has considered a single window treated as two separate parts. The methods employed were the use of a BRS protractor and BRS Nomogram; the BRS daylight tables and inter-reflection formula could have been used in an analogous

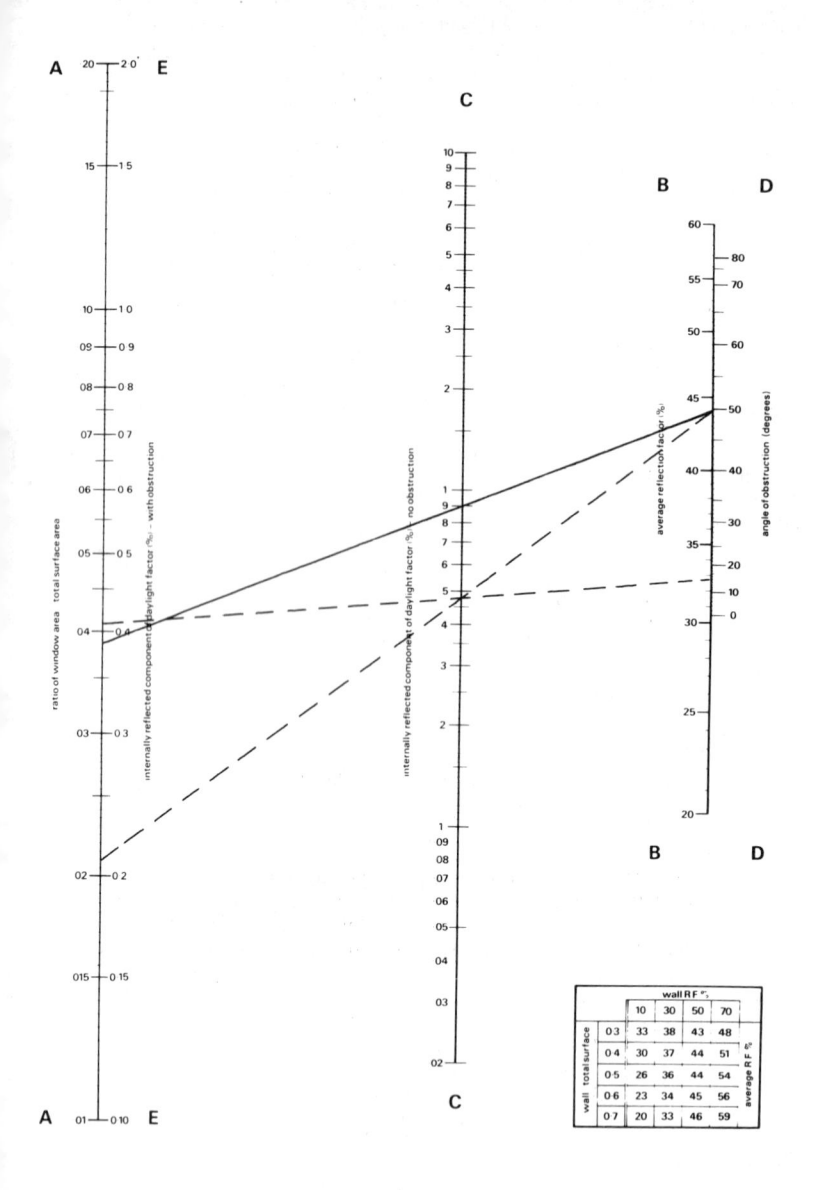

Fig 13 BRS Nomogram 1 for Problem 8 (Reproduced by permission of the Controller, HMSO. Crown copyright)

manner. The general method can be applied to any number of separate windows in a room. The main points to note are:
(i) Sky components and externally reflected components can be determined separately for different windows and then added together.
(ii) When using the BRS protractor for a reference point which is to one side of a window the correction factors for windows subtending less than 180° on plan must be subtracted since both sight lines are to one side of the zero line.
(iii) Internally reflected components can be found separately for any window by any suitable method and then added together.

EXERCISES (answers on page 186)

1 For the following statement select the correct option or options. The luminance at the zenith for a CIE overcast sky is:
   (a) less than that at the horizon;
   (b) equal to that at the horizon;
   (c) greater than that at the horizon;
   (d) three times that at the horizon;
   (e) one-third that at the horizon.

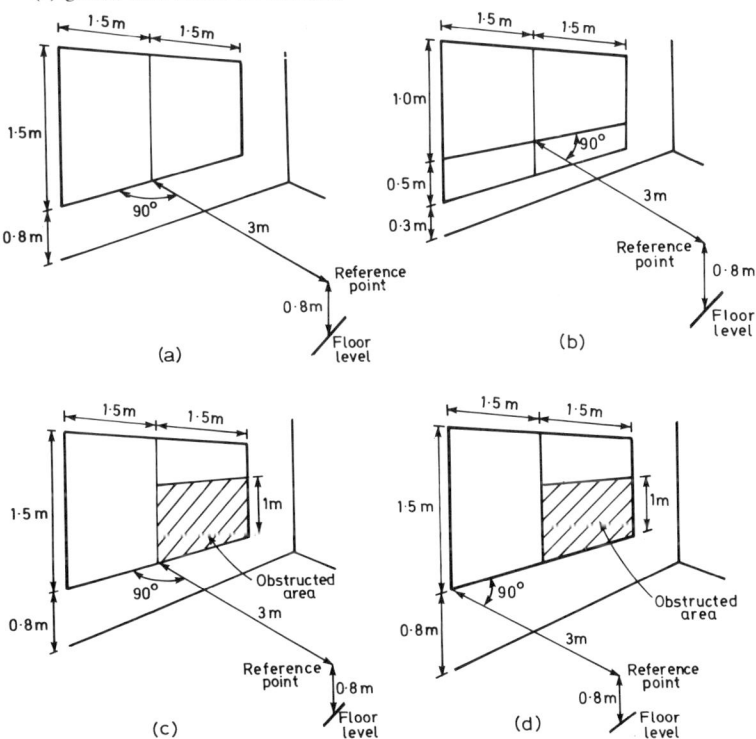

Fig 14 Windows for Exercise 2

2 Using the BRS sky component table find the sky component at the reference point for each of the windows shown in *Figs 14(a), 14(b), 14(c) and 14(d)*. The dimensions shown apply to the outer face of the wall containing the window.

3 Using the BRS sky component table find the externally reflected component at the reference point for the window shown in *Fig 14(c)*.

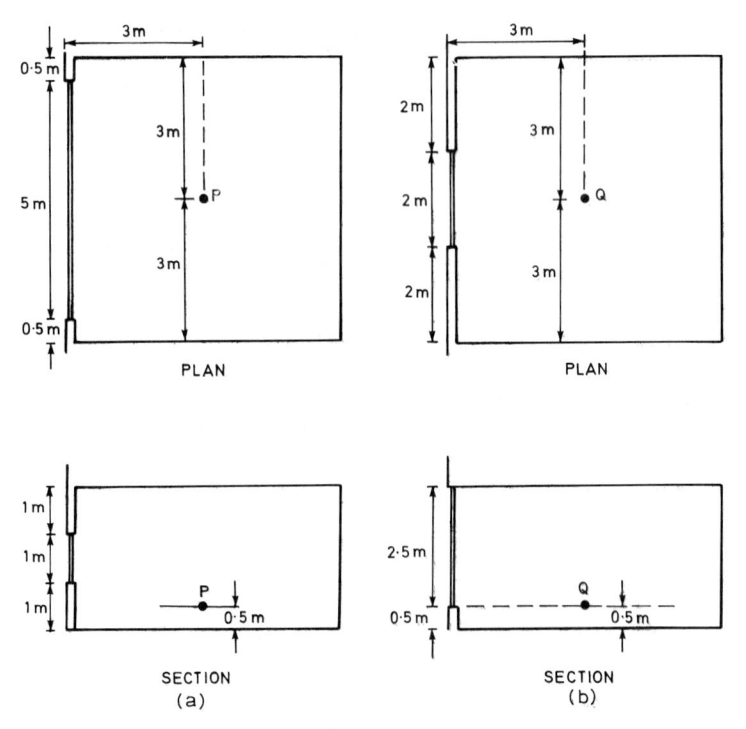

Fig 15 Plans and sections for Exercise 4

4 *Figs 15(a) and 15(b)* show the plans and sections of two rooms of the same size. Both rooms have the same area of glazing but different window shapes. By finding the sky components at the reference points P and Q decide which of the following options are correct:
(a) the sky component at P is approximately 1%;
(b) the sky component at P is approximately 1.9%;
(c) the sky component at Q is approximately 2.7%;
(d) the sky component at Q is approximately 1.3%;
(e) a long low horizontal window gives better penetration of daylight than a tall vertical window;
(f) a tall vertical window gives better penetration of daylight than a long low horizontal window.

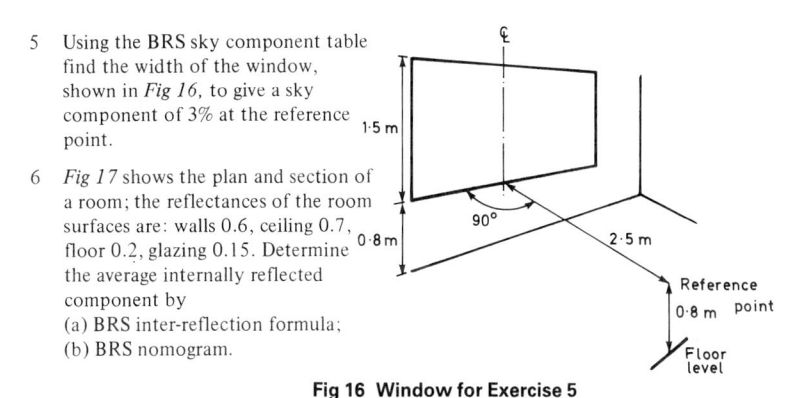

5 Using the BRS sky component table find the width of the window, shown in *Fig 16*, to give a sky component of 3% at the reference point.

6 *Fig 17* shows the plan and section of a room; the reflectances of the room surfaces are: walls 0.6, ceiling 0.7, floor 0.2, glazing 0.15. Determine the average internally reflected component by
(a) BRS inter-reflection formula;
(b) BRS nomogram.

Fig 16 Window for Exercise 5

PLAN

SECTION

Fig 17 Plan and section for Exercise 6

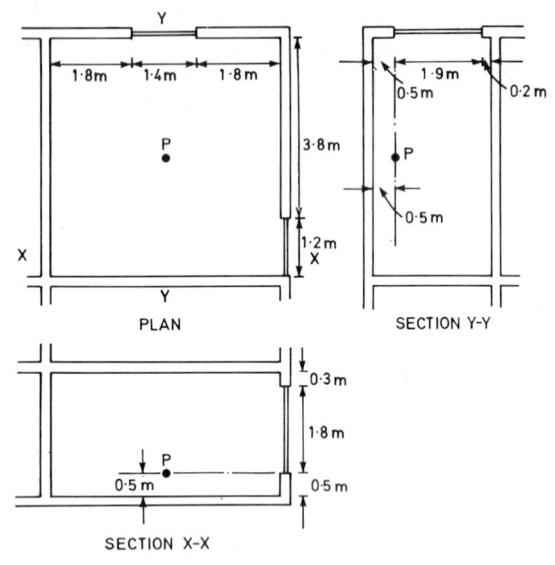

Fig 18 Plan and sections of room for Exercise 7

7. *Fig 18* shows the plan and sections of a room 5 m by 5 m by 2.6 m high. The reference point P is at the centre of the plan and at a height of 0.5 m above floor level. Using BRS protractor No. 2 and Nomogram 1 determine:
   (a) the sky components at P due to each of the windows;
   (b) the average internally reflected components due to each of the windows assuming a wall reflectance of 0.5;
   (c) the daylight factor at P assuming the following correction factors: glazing type 1.0; glazing bars 0.7; dirt on windows 0.9; deterioration of decorations 0.95.

8. *Fig 19* shows the plan and section of a room with four windows. Outside one window is an obstruction which is sufficiently long to obstruct the complete width of the window from the reference point P. The reference point, P, is at the centre of the plan of the room. Using BRS protractor No. 2 and Nomogram 1, and assuming a wall reflectance of 0.6 determine:
   (a) (i) the sky component at P due to an unobstructed window
       (ii) the sky component at P due to the obstructed window
       (iii) the total sky component at P due to the four windows.
   (b) the externally reflected component.
   (c) (i) the average internally reflected component due to an unobstructed window
       (ii) the average internally component due to an obstructed window
       (iii) the total average internally reflected component due to the four windows.
   (d) the daylight factor at P using the following correction factors:
       glazing type 1.0; glazing bars 0.85; dirt on window 0.85; deterioration of decorations 0.9.

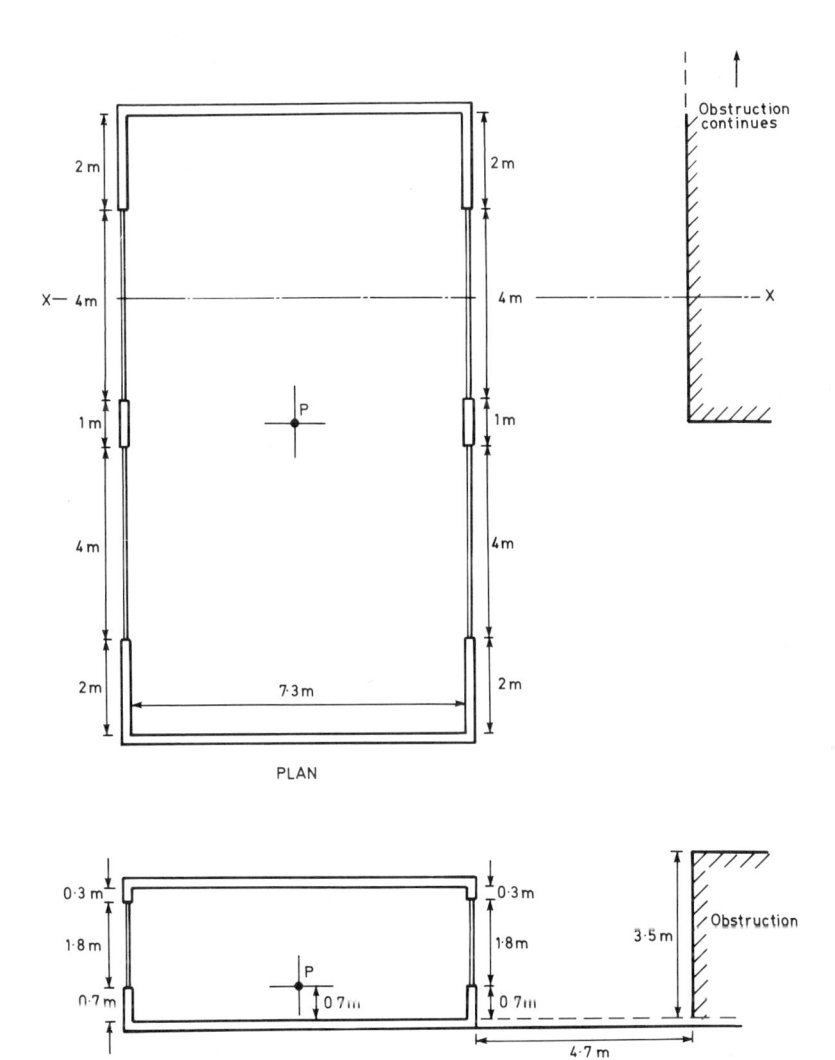

Fig 19 Plan and section of room for Exercise 8

Answers to Exercises

CHAPTER 1 (see page 10)
1(i) 18.4°C;
 (ii) 20°C;
2(i) 1.4448 kPa;
 (ii) 79.5%;
 (iii) 90.4%;
3(i) 1.24 kPa;
 (ii) 0.0077 kg/kg dry air;
 (iii) 77%;
 (iv) 10°C;
4   48.6%;
5   66%, 13.5°C;
6   (a), (c);
7   (c);
8   (b);
9   (d);

CHAPTER 2 (see page 36)
1(a) About 0.94 W/mK;
 (b) About 0.54 W/mK;
2   0.184 W/mK;
3   0.61 W/mK, about 0.11 m²
4(i) 0.15 m²K/W;
 (ii) 0.11 m²K/W;
5   (d);
6   (b), (d);
7   0.56 W/m²K;
8   0.92 W/m²K;
9   0.46 W/m²K;
10 50 mm, 1560 W, 390 W;
11 8051W, walls 25.1%,
    windows 25%,
    floor 6.9%,
    roof 18.1%;
    ventilation 24.9%;

12 2766 W, 18.8°C, 23.6°C;
13 1.37 W/m²K;
14 1.47 W/m²K;
15 20.4%;
16 50%;
17 25% 23.5%, 22%,
    20.4%, 18.8%, 17%;
18 19.38%;
19 48.2%;
20(i) Window 313.6 m²,
    rooflight 75 m²;
 (ii) 388.6 m²;
 (iii) 791 m²;
 (iv) 201 m²;
 (v) Case (i) 0.445 W/m²K;
    Case (ii) 0.378 W/m²K;
    Case (iii) not possible,
    either reduce area of
    glazing or decrease
    U-value of roof;
    Case (iv) 0.365 W/m²K;
21 25.1 litres;
22 Temperatures at interfaces:
    R_{si}/plaster = 18.80°C, plaster/glass
    fibre = 18.17°C, glass fibre/blocks =
    13.16°C, blocks/R_{so} = 2.60°C;
23 Temperatures at interfaces:
    R_{si}/plaster = 16.80°C, plaster/blocks =
    15.32°C, blocks/cavity = 5.68°C,
    cavity/bricks = 2.39°C,
    bricks/R_{so} = 0.10°C;

CHAPTER 3 (see page 48)
1(i) 6.37°C;
 (ii) 47%;

2(i) 7.06°C;
 (ii) 1.29 kPa approximately;
 (iii) Yes;
3 See *Fig 1*;
4 See *Fig 2*;
5 (c);

CHAPTER 4 (see page 58)
1 (a), (c);
2 (b);
3 (c);
4 (c);
5 (b);
6 (b), (c);
7 0.28, 0.52;
8 (b) because (c) and (e);
9(i) 0.72;
 (ii) 0.73;
 (iii) 430 W/m^2;
10 (a), (b), (e);
11 8720 W;

CHAPTER 5 (see page 69)
1 340.3 m/s;
2 0.041 s;
3 56.7, 133.3, 170, 226.7, 283.3 Hz;
4 (a) 0.0159 W/m^2,
 (b) 102 dB;
5 0.01 W;
6 69.8 dB;
7 76.7 dB;
8 7 machines but 6 nearly do it;
9 64.5 dB;
10 (c);
11 (d);
12 (a);
13 (a);

CHAPTER 6 (see page 78)
1 (c); 2 (d);

Distance (m)	2	4	6	8	10	12	14	16	18	20
Sound pressure level (dB)	89.6	85.2	83.5	82.6	82.2	81.9	81.7	81.6	81.6	81.5

4 (b), (d);
5 47 Hz;
6 (i) 1.45 seconds;
 (ii) 1.13 seconds;
7 49 m^2;

8 0.27;
9 (c);
10 (c);
11 (b);
12 (a), (c);

CHAPTER 7 (see page 94)
1 (d);
2 0.001995;
3 32 dB;
4 (a);
5 17, 22, 27, 32, 37, 42 dB;
6 4.9 kg/m^2;
7 (b);
8 (b);
9 30 dB;
10 8.82 m^2;
11 (c), (d);
12 (a) 7.2 m^2; (b) 49 dB;
13 (d);
14 33 dB;
15 48 dB;
16 (c);

CHAPTER 8 (see page 111)

1. (b), (c);
2. (c);
3. (c), (d);
4. (a), (d);
5. (c);
6. 87.3 dBA;
7. 0.25 hours;
8. (a) 14, 21.5; 26.3, 33.5, 36.2, 34.5, 42.2, 28.5 dB; (b) 42 kg/m^2;
9. 30, 37, 42, 44, 44, 46, 43, 31 dB;
10. (b);
11. 68 dBA (approx);
12. 82 dBA;
13. 7.9 m;
14. 80.5 dBA;
15. 82.8 dBA;
16. Yes;
17. 109 dB;

CHAPTER 9 (see page 119)

1. (c);
2. (c);
3. (i) 81 dBA;
   (ii) 71 dBA;
   (iii) 73 dBA;
   (iv) 63 dBA;
4. 52 dB;
5. (c);
6. (i) 2.05 s;
   (ii) 0.38 s;
   (iii) 7 dB approx.

CHAPTER 10 (see page 132)

1. (b), (c), (e);
2. (b), (c), (e);
3. (b), (f);
4. (b);
5. (b);
6. (c);
7. (a) 125.7 cd;
   (b) 1579 lumens;
8. 0.63;
9. (b);
10. (b);
11. (b);
12. (c);
13. (a);
14. (a); (c);
15. (a);
16. (c);
17. (a);
18. (c);
19. (b), (c), (e), (f);
10. (b), (c), (e);
21. (b);
22. 30%;

CHAPTER 11 (see page 153)

1. (d);
2. 320 lux;
3. 228 lux;
4. 170 lux approx.
5. 130 lux approx.
6. 205 lux approx.
7. 144 lux;
8. 500 lux;
9. 470 lux;
10. 505 lux;
11. 480 lux;
12. (a) 18,
    (b) Using 6 × 3 array spacing is satisfactory;
13. 13.6;
14. 14.7;

CHAPTER 12 (see page 179)

1. (c), (d);
2. (a) 1.6%; (b) 0.74%; (c) 1.23%; (d) 1.14%,
3. 0.07%;
4. (b), (c), (f);
5. 3.5 m;
6. (a) 1.9%; (b) 1.8%;
7. (a) 1.8%; 0.56%; (b) 0.6%; 0.48%; (c) 2.1%;
8. (a) (i) 0.9%; (ii) 0.47%; (iii) 3.17%;
   (b) 0.09%;
   (c) (i) 0.63%; (ii) 0.5%; (iii) 2.39%;
   (d) 4%;

Index

Acoustic,
   barriers, 116
   sheds, 115
Air,
   temperature, 1
   velocity, 2
Airborne sound insulation, 81
   measurement, 88
Aircraft noise, 109

British zonal system (BZ), 151
BRS,
   inter-reflection formula, 168
   nomograms, 171
   sky component protractor, 162
   sky component table, 158

Casual gains,
   human bodies, 57
   lighting, 58
Cavity index, 143
Chromaticity coordinates, 128
Coincidence effect, 83
Colour,
   light sources, 127
   rendering, 131
   surfaces, 130
   temperature, 127
Comfort, 8
   zones, 9
Condensation,
   interstitial, 42
   surface, 40
Corrected noise level, 104

Daylight factor,
   components, 156
   definition, 156
   externally reflected component, 166
   internally reflected component, 168
   sky component table, 158
Daylight protractor, 162
Decibel,
   addition and subtraction, 65
   A-weighted, 68
   definition, 64
Dew point temperature, 4
Directivity factor, 72

Electric lamps,
   discharge, 126
   fluorescent, 127
   incandescent, 125
   tungsten halogen, 125
Equivalent continuous sound level, 99
Eye, 121
Eyring's formula, 76

Flanking transmission, 87
Flicker, 153

Glare,
   disability, 147
   discomfort, 147
   index, 148
Glass transmission characteristics, 53

Hearing,
   damage, 98
   loss, 98
   mechanism, 97
   threshold, 64
Heat bridges, 26
Heat exchange,
   by convection, 15
   by radiation, 15
Heat loss rate,
   by ventilation, 21
   during winter heating, 22
   through fabric, 18
Heating fuels, 32
Humidity,
   absolute, 3
   measurement, 5
   relative, 4
   specific, 4

Illuminance,
   array of luminaires, 141
   definition, 123
   linear source, 137
   point source, 134
   scalar, 153
Illumination vector, 153
Impact sound insulation, 90
Inverse square law, 134

L_{eq}, 99
L_{10}, 108
Light, 121
Lighting units, 122
Loudness, 67
Lumen method, 141
Luminance, 123
Luminous,
  flux, 122
  intensity, 122

Maintenance factor, 141
Munsell system, 130

Noise,
  aircraft, 109
  annoyance, 101
  construction sites, 106
  demolition sites, 106
  exposure of employed persons, 99
  rating curves (NR), 102
  rating of industrial, 104
  traffic, 107
Noise control,
  at receiver, 119
  at source, 114
  by acoustic sheds, 115
  by barriers, 116
  by distance, 116
  by machinery enclosures, 115
  indoors, 118
  outdoors, 115
Noise Insulation Regulations, 108

Overcast sky, 156

Pattern staining, 35
Perceived noise level, 109
Polar curve, 135
Psychrometric chart, 6

Reflectance,
  average, 143
  effective, 143
Reverberation time, 74
  optimum, 75
Room,
  acoustic design, 77
  constant, 72
  index, 142

Sabine's formula, 74
Saturation vapour pressure, 2
Solar gain, 50
  methods of reducing, 56

Solar radiation, 50
  components, 51
  total, 52
Sound,
  absorbing materials, 73
  absorption coefficient, 71
  equivalent continuous level, 101
  frequency, 62
  in rooms, 71
  intensity, 63
  intensity level, 64
  level meter, 68
  power, 62
  power level, 65
  pressure, 63
  pressure level, 65
  spectra, 69
  transmission coefficient, 81
  velocity, 61
  wavelength, 62
  waves, 60
Sound insulation,
  airborne, 81
  impact, 90
  rating in buildings, 93
  requirements, 91
Sound level,
  meter, 68
  through partitions, 88
Sound reduction index,
  composite partitions, 85
  definition, 81
  multiple leaf partitions, 87
  single leaf partitions, 82
Speech interference level, 101

Temperature,
  air, 1
  distribution, 33
  dry resultant, 8
  environmental, 9
  equivalent, 8
  globe, 8
  mean radiant, 1
Thermal,
  conductance, 13
  conductivity, 12
  insulation regulations, 27
  resistance, 13
  resistivity, 13
  transmittance, *see* U-value
Thermal resistance,
  air spaces, 17
  definition, 13
  external surface, 16

Thermal resistance (*cont.*)
   internal surface, 16
   total, 18

U-value,
   average, 26
   definition, 18
Utilisation factor, 141

Vapour,
   permeability, 42
   resistance, 43
   resistivity, 42
Vision, 121

Waveforms of sound, 69
Weber-Fechner law, 64
Weighting scales,
   sound level meters, 68
Wien's displacement law, 50

Butterworths Technician Series

Mathematics

Mathematics for Technicians 1
F Tabberer

1978   192 pages   246 × 189 mm
0 408 00326 X   Limp   Illustrated

Mathematics for Technicians 2
F Tabberer

1978   156 pages   246 × 189 mm
0 408 00371 5   Limp   Illustrated

Science

Physical Science for Technicians 1
R McMullan

1978   96 pages   246 × 189 mm
0 408 00332 4   Limp   Illustrated

Building Construction, Civil Engineering, Surveying and Architecture

Building Technology 1
J T Bowyer

1978   96 pages   246 × 189 mm
0 408 00298 0   Limp   Illustrated

Building Technology 2
J T Bowyer

1978   96 pages   246 × 189 mm
0 408 00299 9   Limp   Illustrated

Building Technology 3
J T Bowyer

1980   104 pages   246 × 189 mm
0 408 00411 8   Limp   Illustrated

Civil Engineering Technology 3
B J Fletcher and S A Lavan

1980   96 pages   246 × 189 mm
0 408 00426 6   Limp   Illustrated

Construction Science and Materials 2
D Watkins and J Fincham

1981   192 pages approx   246 × 189 mm
0 408 00488 6   Limp   Illustrated

Site Surveying and Levelling 2
W S Whyte and R E Paul

1981   160 pages approx   246 × 189 mm
0 408 00532 7   Limp   Illustrated

Heating and Hot Water Services for Technicians
K Moss

1978   168 pages   246 × 189 mm
0 408 00300 6   Limp   Illustrated

Electrical, Electronic and Telecommunications Engineering

Electrical Drawing for Technicians 1
F Linsley

1979   96 pages   246 × 189 mm
0 408 00417 7   Limp   Illustrated

Telecommunications Systems for Technicians 1
G L Danielson and R S Walker

1979   112 pages   246 × 189 mm
0 408 00352 9   Limp   Illustrated

Transmission Systems for Technicians 2
G L Danielson and R S Walker

1981   72 pages approx   246 × 189 mm
0 408 00562 9   Limp   Illustrated

Radio Systems for Technicians 2
G L Danielson and R S Walker

1981   96 pages approx   246 × 189 mm
0 408 00561 0   Limp   Illustrated

Radio Systems for Technicians 3
G L Danielson and R S Walker

1982   112 pages approx   246 × 189 mm
0 408 00588 2   Limp   Illustrated

Electrical and Electronic Principles 2
I R Sinclair

1979   96 pages   246 × 189 mm
0 408 00433 9   Limp   Illustrated

Electrical and Electronic Applications 2
D W Tyler

1980   204 pages   246 × 189 mm
0 408 00412 6   Limp   Illustrated

Electronics for Technicians 2
S A Knight

1978   112 pages   246 × 189 mm
0 408 00324 3   Limp   Illustrated

Electronics for Technicians 3
S A Knight

1980   160 pages   246 × 189 mm
0 408 00458 4   Limp   Illustrated

Electrical Principles for Technicians 2
S A Knight

1978   144 pages   246 × 189 mm
0 408 00325 1   Limp   Illustrated

Electrical and Electronic Principles 3
S A Knight

1980   160 pages   246 × 189 mm
0 408 00456 8   Limp   Illustrated

Electrical and Electronic Principles 4/5
S A Knight

1982   176 pages approx   246 × 189 mm
0 408 01109 2   Limp   Illustrated

Mechanical, Production, Marine and Motor Vehicle Engineering

Vehicle Technology 1
M J Nunney

1980   112 pages   246 × 189 mm
0 408 00461 4   Limp   Illustrated

Vehicle Technology 2
M J Nunney

1981   96 pages approx   246 × 189 mm
0 408 00594 7   Limp   Illustrated

Engine Technology 1
M J Nunney

1981   120 pages approx   246 × 189 mm
0 408 00511 4   Limp   Illustrated

Manufacturing Technology 2
P J Harris

1979   96 pages   246 × 189 mm
0 408 00410 X   Limp   Illustrated

Manufacturing Technology 3
P J Harris

1981   104 pages approx   246 × 189 mm
0 408 00493 2   Limp   Illustrated

Fabrication, Welding and Metal Joining Processes − A textbook for Technicians and Craftsmen
C Flood

1981   160 pages approx   246 × 189 mm
0 408 00448 7   Limp   Illustrated

Materials Technology for Technicians 2
W Bolton

1981   128 pages approx   246 × 189 mm
0 408 01117 3   Limp   Illustrated

Materials Technology for Technicians 3
W Bolton

1982   128 pages approx   246 × 189 mm
0 408 01116 5   Limp   Illustrated

Materials Technology 4
W Bolton

1981   128 pages approx   246 × 189 mm
0 408 00584 X   Limp   Illustrated

Mechanical Science for Technicians 3
W Bolton

1980   128 pages   246 × 189 mm
0 408 00486 X   Limp   Illustrated

Mechanical Science for Higher Technicians 4/5
D H Bacon and R C Stephens

1981   256 pages approx   234 × 156 mm
0 408 00570 X   Limp   Illustrated

Thermodynamics for Technicians 3/4
D H Bacon and R C Stephens

1982   96 pages approx   234 × 156 mm
0 408 01114 9   Limp   Illustrated

Engineering Instrumentation and Control
W Bolton

1980   144 pages   246 × 189 mm
0 408 00462 2   Limp   Illustrated